SpringerBriefs in Water Science and Technology

SpringerBriefs in Water Science and Technology present concise summaries of cutting-edge research and practical applications. The series focuses on interdisciplinary research bridging between science, engineering applications and management aspects of water. Featuring compact volumes of 50 to 125 pages (approx. 20,000–70,000 words), the series covers a wide range of content from professional to academic such as:

- Timely reports of state-of-the art analytical techniques
- Literature reviews
- In-depth case studies
- Bridges between new research results
- Snapshots of hot and/or emerging topics

Topics covered are for example the movement, distribution and quality of freshwater; water resources; the quality and pollution of water and its influence on health; and the water industry including drinking water, wastewater, and desalination services and technologies.

Both solicited and unsolicited manuscripts are considered for publication in this series.

More information about this series at http://www.springer.com/series/11214

M. Dinesh Kumar · Cecilia Tortajada

Assessing Wastewater
Management in India

M. Dinesh Kumar
Institute for Resource Analysis and Policy
Hyderabad, Telangana, India

Cecilia Tortajada
Institute of Water Policy
Lee Kuan Yew School of Public Policy
National University of Singapore
Singapore, Singapore

ISSN 2194-7244 ISSN 2194-7252 (electronic)
SpringerBriefs in Water Science and Technology
ISBN 978-981-15-2395-3 ISBN 978-981-15-2396-0 (eBook)
https://doi.org/10.1007/978-981-15-2396-0

This Springer imprint is published by the registered company Springer Nature Singapore Pte Ltd.
The registered company address is: 152 Beach Road, #21-01/04 Gateway East, Singapore 189721, Singapore

Preface and Acknowledgements

While concerns over environmental degradation, particularly water pollution, are growing in India, the general perception of the public is that the pollution problems are on the rise and already very serious due to indiscriminate disposal of domestic sewage and industrial effluents in water bodies such as rivers and lakes of the country. However, data emerging from monitoring, of water bodies in India including rivers, lakes, tanks and groundwater, carried out by the national and state level institutions concerned with pollution control do not conform to this perception. Instead, they suggest gradual improvement in the situation at least from a macro perspective, though there are evidence of high degree of pollution in certain stretches of some important rivers.

One major driver of change in the overall water quality management scenario in India over the past 15 years or so is the gradual increase in capacity of wastewater treatment plants, mainly in metropolitan cities, and Class I and Class II cities/towns. Currently, around 38% of the wastewater generated in the country is being treated —a remarkable improvement from the situation that existed around 15 years ago. Some new wastewater treatment technologies that are better suited to the climatic and socioeconomic conditions that prevail in India are being introduced. The old plants are being upgraded. This is the result of changing legal, institutional and policy environment with regard to water quality management and pollution control. Water quality testing to monitor the performance of wastewater treatment plants is also being undertaken at major plant sites. But it is important to know how far the effluent quality norms are being adhered to by the agencies that run the plants and the cost of running these plants.

That said, in the future, more cities and towns of India will be investing in wastewater treatment, especially in the water-scarce regions, as the environmental quality standards are made more stringent. More importantly, in the medium and long term, many large cities from water-scarce regions would witness investments for tertiary treatment of wastewater for municipal water supplies, as freshwater becomes increasingly scarce and cost of tapping new sources becomes high. Therefore, the quantum of treated wastewater, available as a new source of water for meeting irrigation and municipal water supply needs, is likely to increase substantially. In such a scenario, it is important to know the scale of investment that cities would witness for this. However, there is not much long-term planning with regard to investments and returns from wastewater treatment and reuse of treated wastewater. Government policies with regard to reuse of treated water are nonexistent in most Indian states.

Economic analysis of wastewater treatment plants that can be used in the country's context is largely absent. Very little is known about the potential trade-off between improving the environmental performance of the treatment plants (quality of treated effluent) and their economics, particularly the cost of the treatment systems and the (social, environmental and economic) benefits. Given the fact that urban water utilities are showing resistance to charge the consumers to recover the treatment costs, and the willingness on the part of the communities living in and around urban areas to pay for ecosystem services is poor, reuse potential of treated wastewater (for municipal, irrigation, and industrial water supply) and the economic gains from the same need to be explored. In that case, it is important to know the size of the market for treated wastewater of different quality, say, irrigation and drinking water supplies and where such markets are likely to emerge.

The monograph provides a detailed analysis of the situation with respect to: water quality status of selected rivers; the existing legal, institutional and policy environment with regard to water quality management and pollution control; the state of wastewater generation, collection, treatment and reuse in cities and towns; and the performance of treatment plants. It also attempts to address some of these questions in a comprehensive way.

Finally, this research was supported by the Institute of Water Policy, Lee Kuan Yew School of Public Policy, National University of Singapore.

Hyderabad, India   M. Dinesh Kumar
Singapore   Cecilia Tortajada

Contents

1 Introduction   1
   1.1 Context   1
   1.2 Objectives of the Study   4
   1.3 Methodology   4

**2 The Legal, Institutional and Policy Regime for the Control
of Water Pollution in India**   7
   2.1 Introduction   7
   2.2 Water Quality Monitoring in India   8
   2.3 Legal and Institutional Regime for Water Quality
Management   8
   2.4 Issues Related to Pollution Control   12

**3 Status of Water Quality in India and Compliance with Pollution
Control Norms**   13
   3.1 An Overview of Water Quality Status in India   13
   3.2 Compliance with Pollution Control Norms   14
   3.3 Findings and Conclusions   16

4 Effectiveness of Wastewater Collection and Treatment Systems   17
   4.1 Wastewater Generation and Collection in India's
Urban Areas   17
   4.2 Sewage Treatment Capacity   18
   4.3 Sanitation Access in Urban India   21
   4.4 Findings and Conclusion   22

5 Health Impacts of Water Pollution and Contamination   23
   5.1 Overview of Public Health Impacts of Water Contamination
in India   23

5.2   Case Studies on Water Pollution and Impacts
       on Public Health   24
5.3   Findings and Conclusions   30

6   Reuse of Treated Wastewater: Present Scenario   31
6.1   Introduction   31
6.2   Cases of Reuse of Treated Wastewater   31
6.3   Concluding Remarks   33

7   Wastewater Treatment Technologies and Costs   35
7.1   The Context   35
7.2   Conventional or Centralized Wastewater Treatment   36
7.3   Treatment Technologies Used in Sewage Treatment Plants:
       Performance and Costs   37
7.4   Decentralized Systems for Wastewater Treatment:
       Performance and Costs   39
7.5   Findings and Conclusion   41

**8   Case Studies on Performance of Wastewater Treatment
       Systems**   43
8.1   Introduction   43
8.2   General Profile of the Plants and Locations   43
8.3   Working of the Wastewater Treatment Plants   45
8.4   Findings and Conclusion   48

9   Environmental Sustainability and Economic Viability   49
9.1   Introduction   49
9.2   Environmental Sustainability Versus Economic Viability   50
9.3   Findings and Conclusion   52

10   Growth of Treatment Plants and Reuse of Treated Wastewater . . .   53
10.1   Introduction   53
10.2   Future Growth of Treatment Plants and Wastewater Reuse   53
10.3   Future Investments for Wastewater Treatment   58
10.4   Findings and Conclusion   58

11   Market for Treated Wastewater in India   59
11.1   Introduction   59
11.2   Future Market for Treated Wastewater   60
11.3   Findings and Conclusion   61

12   Conclusions and Areas for Future Research   65
12.1   Summary of Findings and Conclusions   65
12.2   Future Research Areas on Wastewater   66

Contents   ix

Appendix 69

References 77

Index 83

List of Figures

Fig. 4.1   Water supply and wastewater generation in different
classes of cities   18
Fig. 4.2   Water supply and wastewater treatment scenarios in different
classes of cities   19
Fig. 4.3   Total sewage generation and STP capacity in Indian states   20
Fig. 4.4   Statewise sewage treatment capacity and volume of wastewater
treated, by state   20
Fig. 7.1   Performance of different wastewater treatment technologies
in BOD removal   37
Fig. 7.2   Performance of different wastewater treatment technologies
in COD removal   38
Fig. 10.1 Projected additional capacity requirements for wastewater
treatment systems (MLD)   57

List of Tables

Table 2.1   Laws, rules, manuals, policies, guidelines and reports on sanitation, waste management and pollution control in India 11

Table 5.1   Water-borne diseases: bacterial, viral and protozoan 25

Table 5.2   Water-related diseases: vector-borne 27

Table 5.3   Reported cases (in numbers) of water-borne diseases and deaths in India, 2013–2017 28

Table 7.1   Conventional technologies used in sewage treatment plants in India (number of plants) 38

Table 7.2   Cost comparison of various technologies used in sewage treatment plants 39

Table 7.3   Cost comparison of different decentralized sewage treatment technologies 41

Table 8.1   Features and key performance indicators of the wastewater treatment plants 44

Table 8.2   Details of plant operation, number of staff engaged, annual operation and maintenance charges, and electricity charges 46

Table 9.1   Cost of pollution abatement in terms of BOD removal for different BOD concentrations (Rs./KL) 51

Table 10.1   Projected trajectory of growth in wastewater infrastructure in Indian cities and towns 55

Table 10.2   Projected future scale of investments in two types of wastewater treatment systems in different categories of cities 57

Table 11.1   Projected demand for treated wastewater in different categories of cities 62

Table A.1   Water quality of the Musi River at Hyderabad 70

Table A.2   Water quality of the Godavari River at Nashik,
             Maharashtra   71
Table A.3   Water quality of the Yamuna River at Delhi   72
Table A.4   Water quality of the Yamuna River at Mathura   73
Table A.5   Water quality of the Yamuna River at Agra   74
Table A.6   Water quality of the Ganga River at Varanasi   75

Chapter 1
Introduction

Abstract This chapter provides an overview of the water management problems facing India due to growing demand-supply gap, and how water quality deterioration due to increasing pollution of surface water and groundwater from untreated or partially treated industrial effluents, urban sewage and poor sanitation compound these challenges, while posing a serious threat to public health. The chapter also outlines the objectives of the study which forms the basis for this monograph, the scope and the methodologies used.

Keywords India · Water demand · Water pollution · Industrialization · Effluent · Water treatment

1.1 Context

The demand for water for various competitive uses and the environment is on the rise in India (Kumar 2010). The Standing Subcommittee of the Ministry of Water Resources (2000) estimates that the total water demand across all sectors in India will rise from 813 billion cubic metres (BCM) in 2010 to 1093 BCM in 2025 and 1447 BCM in 2050. An assessment by the International Water Management Institute (Amarasinghe et al. 2008) considering the business-as-usual scenario puts these figures at a maximum of 680, 833 and 900 BCM, respectively (GOI 1999). While the usable water resources in India were estimated at 1123 BCM in 2015, the availability is not uniform across the country (Ministry of Statistics and Programme Implementation 2016a). In addition to the spatio-temporal variations in availability, vagaries of the monsoon and the predicted variability of climate in the future, human, development and management challenges pose the greatest threats to water availability (Source: Water Resources Information System of India, website accessed in 2019). In many parts of India, extraction of large volumes of groundwater, far exceeding the natural replenishment, has led to a steep decline in water tables.

Water security is the availability of an acceptable quantity and quality of water for health, livelihoods, ecosystems and production, coupled with an acceptable level of water-related risks to people, environments and economies. Achieving water security is the only way for a poor country to grow and lift its people out of poverty (Grey and Sadoff 2007). A condition of water stress is said to exist when there is not enough water for all uses, whether agricultural, industrial or domestic. Factors that contribute to water stress are excessive withdrawal from surface and underground sources, inefficient use of water, and pollution of freshwater bodies.

The Central Ground Water Board (2014) estimates total annual groundwater draft in India at only 22.72 BCM for domestic and industrial purposes, against 222.36 BCM for irrigation. But many Indian cities have to rely on imported water from distant sources, as the local aquifers are not capable of meeting the concentrated water demand due to poor yields. Cities with larger populations rely more on surface water than on groundwater sources, and this leads to conflicts with other uses too. In cities where groundwater is the source of water supply, aquifers are being depleted as excessive pumping occurring within small geographical areas creates cones of depression (Mukherjee et al. 2010).

The situation is made worse by the pollution of water bodies. Discharge of untreated sewage into water bodies—both surface and groundwater—may be responsible for polluting three-fourths of surface water resources; the volumetric extent of surface water pollution in India may be as high as 80%. Unfortunately, environmental sanitation has not kept pace with water supply in India, and as a result almost 80% of the water supplied for domestic use comes back as wastewater to pollute freshwater and soil. In rural areas, open defecation in the catchments of rivers, lakes and ponds pollutes surface water bodies, and in certain areas poorly designed on-site sanitation systems cause groundwater pollution. In urban areas, through sewerage systems are usually present, untreated or partially treated sewage flows into natural bodies such as streams, rivers, lakes and ponds from cities and towns where wastewater treatment plants are either absent or have inadequate capacity. It is estimated that 75–80% of water pollution (in terms of volume of water polluted) is from domestic sewerage (WaterAid 2016a).

Fast industrialization and rising water demand for energy production are other factors causing increasing pollution of natural water systems. The World Bank has estimated that the water demand for industrial uses and energy production in India will grow at 4.2% annually (IDFC 2011). The wastewater from industrial units further pollutes freshwater when discharged into water bodies, as does agricultural runoff. As the water table sinks due to mining of aquifers for urban water supply, the groundwater may become degraded do to seepage from polluted surface water bodies. Thus, the amount of freshwater available for various uses is on the decline due to pollution and contamination from domestic, industrial and agricultural sources.

Most cities in India are struggling to supply water to their residents (Kumar 2014). The situation is no better in rural areas. Rural households already spend a considerable part of their limited incomes on acquiring clean drinking water, often having to tap a range of different schemes running in their villages, in addition to private provisions like investing in bore wells, storage tanks and so on (World Bank 2008). At the same time, sewage treatment facilities are far from adequate. But the situation is more critical in urban areas due to very high population density. The pollution of rivers in cities also affects the rural areas downstream. The public health cost of diseases caused by consumption of polluted water (due to polluted source water or contamination of the treated water during transmission) and poor sanitation conditions is very high (Tripathi 2018).

All these observations show that treatment and reuse of wastewater is an absolute necessity, not only to reduce water demand but also to prevent pollution of freshwater resources. While it is important to find ways to reduce demand, by increasing water use efficiency in the agricultural, industrial and domestic sectors and reducing transmission losses and pilferage, treating the wastewater for reuse for different purposes and reducing the pollution of our water bodies is equally important, so long as it makes good economic sense from the point of view of full costs and benefits (Kumar 2014).

An integrated approach to water management that links water resources management across different sectors and interest groups, from local to international scales (Global Water Partnership and International Network of Basin Organizations 2009), is necessary to manage freshwater, wastewater and stormwater to secure reliable and sustainable water supply for the short and long term. Alternatives like water reuse are also to be pursued. As suggested by Esposito et al. (2005), decentralized wastewater treatment, with either satellite or cluster systems, enables more flexible and cost-effective treatment configurations with innovative technologies to reclaim water. These systems have reduced costs (which include costs associated with transmission to reuse sites and necessary retrofits) as well as reduced risk of effluent toxicity from mixed wastewater sources.

Treatment and reuse of wastewater has a major role in achieving water security and environmental protection. The Swachh Bharat Abhiyan (referred to as Clean India Mission) will be successful only if water is available in the toilets constructed as part of the project, and the wastewater generated is treated, and the treated water is reused or disposed-off in a sustainable manner.

Given this background, a study was undertaken to: make a quick assessment of wastewater treatment and reuse situation in India comprising the legal, institutional and policy frameworks, types of treatment systems in use and their state-wise distribution, their capacities, operational efficiency, cost of treatment, the institutional set up for running the plants; and project the future growth potential of wastewater treatment capacity, including the scale of investments required, and the market for treated wastewater.

1.2   Objectives of the Study

The objectives of the study were:

1. To provide a comprehensive analysis of the various institutional, policy, legal and regulatory frameworks related to water quality protection and management;
2. To assess

   a. Levels of investments in wastewater treatment systems,
   b. Effectiveness of implementation of wastewater treatment systems in the urban, industrial and agricultural sectors,
   c. Institutional capacities for their operation and maintenance,
   d. Extent of compliance of pollution control and environmental protection norms,
   e. Environmental and health implications of discharge of raw or non-properly treated wastewater to the environment, and
   f. Potential areas of reuse of treated wastewater, including sectors, regions and the quantum of water;

3. And to project

   a. The economic benefits and costs associated with reuse in those sectors, and
   b. The future uses of treated wastewater, including preliminary assessment of benefits and costs.

The scope of work included literature review; collection of primary data and information and discussions with the officials in-charge of operating and maintaining Sewage Treatment Plants, including in some large cities in India. For the first three items under objective 2, data were obtained for case study locations only, using primary survey.

1.3   Methodology

For this monograph, we analyzed time series data published by the Central Pollution Control Board (CPCB) on the status of water quality of aquatic resources in India the reports from 1999 to 2013, and datasets of 2014 (source: https://cpcb.nic.in/NWMP-data-2014), and inventories (CPCB 2005, 2015) of sewage treatment plants in Indian cities. We also reviewed a report by the CPCB (2013b) on the performance of 152 sewage treatment plants across India; a report by the Central Public Health and Environmental Engineering Research Organisation (2005) on the status of water supply, sewerage collection and disposal and wastewater treatment in different categories of cities in India; international scientific literature on the economics of wastewater treatment and reuse, and national literature on water pollution, wastewater reuse and health impacts of water pollution in India; the CPCB (2008) guidelines on water quality management in India; a report by Central Ground Water Board (2014) on the status of groundwater resources; and the book and articles by the first author based on research on integrated urban water management in India (Kumar 2010, 2014, 2018). All these reports are in the public domain.

For the study, we have visited wastewater treatment plants in three cities and spoke with senior officials who are in charge of the plant operations.

Chapter 2
The Legal, Institutional and Policy Regime for the Control of Water Pollution in India

Abstract This chapter discusses the range of legal measures, institutional and policy interventions and major programmes for water quality management and pollution control in India, including that for water quality monitoring. It lists all acts, government rules, policies, manuals and guidelines on sanitation, waste management and pollution control in India. It also analyses how effective these measures were in controlling pollution of India's water resources, and discusses the issues involved in pollution control.

Keywords Water quality monitoring · Water quality management · Municipal solid waste · Biological oxygen demand · National river conservation directorate · National lake conservation plan

2.1 Introduction

Preventing water pollution would require bringing about behavioural change on two fronts: effluent generation; and effluent disposal (WWAP 2017). Unlike many developed countries in Europe, the Americas, Asia and Oceania, India has not started using market instruments, such as sewerage tax or pollution tax, to control wastewater generation and pollution. In some large cities, a sewerage cess is collected by the water utility or the municipal corporation that partly covers the operation and maintenance costs of the sewerage system. The instrument it has tried so far is a regulation supported by the Water Pollution (Prevention and Control) Act of 1974 (amended 1988).

The act aims to prevent and control water pollution and to maintain and restore the wholesomeness of water by establishing Central and State Pollution Control Boards (the CPCB and SPCBs) to monitor and enforce the regulations. The act defines the composition of the boards and the terms and conditions of service of their members. The Water (Prevention and Control of Pollution) Cess Act of 1977 (amended 2003) provides for the levy and collection of a cess on water consumed by persons operating and carrying on certain types of industrial activities to augment the resources of the CPCB and SPCBs. This act sets effluent standards and penalties for non-compliance for effluent-discharging bodies.

The CPCB advises the government on any matter concerning the prevention and control of water pollution, coordinates pollution control activities and provides technical assistance and guidance. The CPCB and SPCBs collect, compile and publish technical and statistical data relating to water pollution and the measures devised for its effective prevention and control. They prepare manuals, codes and guidelines relating to the treatment and disposal of sewage and trade effluents and disseminate information related to the same.

2.2   Water Quality Monitoring in India

Water quality monitoring is a prerequisite for assessing the status of maintenance and restoration of water bodies as well as the extent of pollution. Water quality monitoring is performed by the CPCB and the SPCBs with the following objectives:

- Rational planning of pollution control strategies and their prioritization
- Assessing the nature and extent of pollution control needed in different water bodies
- Evaluating the effectiveness of pollution control measures already in place
- Analyzing water quality trends over time
- Assessing the assimilative capacity of a water body to make pollution control cost-effective
- Understanding the environmental fate of different pollutants
- Assessing the suitability of water for different uses.

As of 2012, the water quality monitoring network in India had 2500 stations in 28 states and six Union Territories, covering 445 rivers, 154 lakes, 12 tanks, 78 ponds, 41 creeks and seawater channels, 25 canals, 45 drains, 10 water treatment plants (raw water) and 807 wells. Of the 2500 stations, 1275 were on rivers, 190 on lakes, 45 on drains, 41 on canals, 12 on tanks, 41 on creeks and seawater channels, 79 on ponds and 10 on water treatment plants (raw water), and 807 were groundwater stations (CPCB 2013a). The inland water quality monitoring network is operated under a three-tier programme: the Global Environmental Monitoring System, the Monitoring of Indian National Aquatic Resources System and the Yamuna Action Plan (CPCB 2013a). In addition to general parameters, and core parameters such as pH, dissolved oxygen (DO) and biochemical oxygen demand (BOD), biological monitoring and monitoring of trace metals and pesticides are undertaken (CPCB 2013a).

2.3   Legal and Institutional Regime for Water Quality Management

A CPCB water quality management plan covers setting water quality goals; monitoring water quality; identifying the nature and magnitude of pollution; inventorying the sources of pollution; collating water quantity information; selecting technologies for pollution control; financing waste management; maintaining sewage treatment plants; and controlling industrial pollution, including recycling and resource recovery, use of clean technologies, and setting wastewater discharge standards and charges for residual pollution (CPCB 2008).

The CPCB has developed a concept of 'designated best use': of the several uses of water of a particular body, the use which demands the highest quality is its designated best use (http://www.cpcb.nic.in/Water_Quality_Criteria.php). This classification helps water quality managers and planners set water quality targets and design suitable restoration programmes for various water bodies. The five designated best uses are:

- drinking water source without conventional treatment but after disinfection
- outdoor bathing (organized)
- drinking water source after conventional treatment and disinfection
- propagation of wildlife and fisheries
- irrigation, industrial cooling, controlled waste disposal.

The CPCB suggests that a major part of the cost of waste management should be borne by the urban population, according to the 'polluter pays' principle, which can be applied to domestic and industrial dischargers to induce waste reduction and treatment and can provide a source of revenue to finance investments in wastewater treatment.

To protect the water quality of rivers, the National River Conservation Directorate (NRCD) was established by the government of India under the Ministry of Environment, Forest and Climate Change (MoEF) to provide technical and financial support to state governments for development of the sewage treatment capacity of municipalities that are discharging their wastewater into natural water bodies. The river cleaning programme in the country was initiated with the launch of the Ganga Action Plan in 1985 (Ministry of Environment, Forest and Climate Change 2006) and was expanded to cover other rivers under the National River Conservation Plan in 1995 (Bassi et al. 2014). The pollution abatement works are implemented on a cost-sharing basis between the central and state governments and include collection, transportation and treatment of municipal sewage, river-front development, low-cost sanitation, and electric crematoria, while prevention and control of industrial pollution is addressed by the CPCB and SPCBs' Pollution Control Committee.

The NRCD's National Lake Conservation Plan 1993 has a 70:30 funding pattern and includes the core components of interception, diversion and treatment of wastewater before its entry into lakes; catchment area treatment; shoreline protection; and in lake treatment, while non-core activities include lakefront eco-development and public participation (Ministry of Environment, Forest and Climate Change 2007) . The National Wetlands Conservation Programme promotes the conservation of wetlands, including Ramsar sites, in the country (Bassi et al. 2014).

The vision of the National Urban Sanitation Policy 2008 is that 'all Indian cities and towns become totally sanitized, healthy and liveable and ensure and sustain good public health and environmental outcomes for all citizens with a special focus on hygienic and affordable sanitation facilities for the urban poor and women' (Ministry of Urban Development 2008).

The Municipal Solid Wastes (Management and Handling) rules were introduced in 2000 by the MoEF, requiring urban local bodies to collect waste in a segregated manner and to transport, process and dispose it using safe and scientific methods (Ministry of Environment, Forest and Climate Change 2000) . This was replaced in 2016 by the new Solid Waste Management Rules (Ministry of Environment, Forest and Climate Change 2016a). This was the sixth in a series of waste management rules notified by MoEF, the other five covering plastic, e-waste, biomedical, hazardous, and construction and demolition waste. In 2016, the Central Public Health and Environmental Engineering Organization (CPHEEO), in collaboration with Deutsche Gesellschaft für Internationale Zusammenarbeit (GIZ), had also published a Municipal Solid Waste Management Manual, with the aim of guiding all urban areas in the country towards sustainable solid waste management, adopting waste minimization at the source with an emphasis on the 3 R's of reduce, reuse and recycle (CPHEEO and GIZ 2016). The manual deals in detail with all aspects of municipal solid waste management, including planning, institutional, financial and technical, providing a step-by-step guide.

The 2008 National Urban Sanitation Policy made local governments responsible for behavioural change, total sanitation and 100% safe waste disposal. It envisages that cities will implement city sanitation plans, prioritizing areas that need urgent attention and implementing long-term plans in parallel, with emphasis on mobilizing all city stakeholders and the importance of behaviour change, practices and installations for safe, sanitary and sustainable disposal of all wastes of the city. It is the responsibility of the state governments to draft state urban sanitation policies, under which the cities can develop their own sanitation strategies (Ministry of Urban Development 2008). The Swachh Bharat Abhiyan also recommends that the state-level High-Powered Committee, formed to authorize institutes for the technical and economic appraisal of Detailed Project Reports for projects recommended by ULBs, facilitate preparation of a state sanitation strategy and city sanitation plans, as envisaged under the 2008 National Urban Sanitation Policy, by the State Mission Directorate (Government of India 2017). While this is a step in the right direction, allowing cities to develop their own strategies to best suit local conditions, there has not been noticeable change on the ground.

The Swachh Survekshan (Urban Sanitation) report (Ministry of Urban Development 2017b) presents the results of a massive sanitation survey undertaken across 500 cities in India. It was intended to provide a comprehensive assessment of the sanitation status of cities and foster a spirit of competition among cities by publishing their scores and rankings. Although the report considers progress in the construction of individual and community toilets and the presence of a processing plant for municipal solid waste, it does not consider wastewater treatment or septage/faecal sludge management.

The draft National Urban Faecal Sludge and Septage Management Policy (Ministry of Urban Development 2017a) says that the problem of faecal sludge and septage/sewerage must be addressed in a holistic manner, with a strategy that provides for minimum needs and is appropriate and affordable for all areas and population considering the local situation. The policy calls for enabling provisions in the form of suitable regulation and institutional framework, capacity building, and education and awareness among all stakeholders. The policy also points to the need to ensure the efficiency of systems in place for on-site sanitation. The faecal sludge output needs to be managed in an environmentally safe manner, including the proper engineering design, construction and maintenance of septic tank systems, pit latrines and other systems generating faecal sludge.

Table 2.1 Laws, rules, manuals, policies, guidelines and reports on sanitation, waste management and pollution control in India

1974: Water (Prevention and Control of Pollution) Act (Government of India 1974)
1977: Water (Prevention and Control of Pollution) Cess Act (Government of India 1977)
1988: Water (Prevention and Control of Pollution) Act, Amendment (Government of India 1988)
2000: Municipal Solid Wastes (Management and Handling) Rules (Ministry of Environment, Forest and Climate Change, 2000)
2003: Water (Prevention and Control of Pollution) Cess Act, Amendment (Government of India 2003)
2008: National Urban Sanitation Policy (Ministry of Urban Development 2008)
2008: Guidelines for Water Quality Management (Central Pollution Control Board 2008)
2011: Water Pollution in India, Report No. 21 of 2011–12 (Comptroller and Auditor General of India 2011)
2012: Status of Water Quality in India, Monitoring of Indian National Aquatic Resources Series MINARS/36/2013-14 (Central Pollution Control Board 2013a)
2016: Municipal Solid Waste Management Manual (Central Public Health and Environmental Engineering Organization and Deutsche Gesellschaft für Internationale Zusammenarbeit 2016)
2016: Plastic Waste Management Rules (Ministry of Environment, Forest and Climate Change, 2016b)
2016: E-Waste (Management) Rules (Ministry of Environment, Forest and Climate Change, 2016c)
2016: Bio-medical Waste Management Rules (Ministry of Environment, Forest and Climate Change, 2016d)
2016: Construction and Demolition Waste Management Rules (Ministry of Environment, Forest and Climate Change, 2016e)
2016: Hazardous and Other Wastes (Management and Trans-boundary Movement) Rules (Ministry of Environment, Forest and Climate Change, 2016f)
2016: Solid Waste Management Rules (Ministry of Environment, Forest and Climate Change 2016a)
2016: Swachhta Status Report (Ministry of Statistics and Programme Implementation 2016b)
2017: Swachh Survekshan (Urban Sanitation Report) (Ministry of Urban Development 2017b)
2017: Draft Policy on National Urban Faecal Sludge and Septage Management (Ministry of Urban Development 2017a)

Table 2.1 lists the laws, rules, manuals, policies, guidelines and reports available in India to help analyze, monitor and combat water pollution and protect water resources.

2.4   Issues Related to Pollution Control

In spite of all the institutional and legal support for the control of water pollution across the country, an audit by the Comptroller and Auditor General of India (2011) found that there is no specific policy on water pollution at either the central or state level incorporating prevention of pollution, treatment of polluted water and ecological restoration of polluted water bodies, and that without this, government efforts in these areas would not get the required emphasis and thrust. Neither does the Water (Prevention and Control of Pollution) Act of 1974 address the concern of restoration of polluted water bodies. The report found that 'the low quantum of penalty and the failure of the State in enforcing the provisions of the Act strictly to secure prevention and control of water pollution have led to the situation where the cost of non-compliance became significantly lower than the cost of compliance'. It also found that the 'highly tolerant inspection regime of the State Pollution Control Boards ensures that the cost of defiance, non-adherence and violations are lower than the cost of compliance'.

There are also problems with the institutional set-up for pollution control. First, the SPCBs have very limited human resources to monitor water quality (with a reasonable frequency and geographical spread of collection, testing and interpretation) in all the water bodies that either constitute the source for water supply schemes or serve the ecosystem, and to disseminate the results to the concerned parties. Many labs are ill-equipped to measure organic pollutant parameters, and in many cases the samples are tested after a long holding period, which alters the parameters (Seth 2011). The inadequate number of monitoring stations and the fact that many of them are not located in cities or immediately downstream of the points of pollution are also some of the reasons for the lack of a realistic picture of the extent of water pollution in the country. Hence, the data seem to paint a picture far rosier than the reality.

Second, while the SPCBs themselves do not have the legal power to penalize polluters, they are required to pursue legal action against them. But they have limited organizational capabilities to pursue legal action against violators of pollution control norms or those who do not comply with effluent disposal standards, with delays in judicial processes. Under such circumstances, there is a tendency to interpret the results of water quality monitoring in such a way as to downplay the magnitude of the problems.

Chapter 3
Status of Water Quality in India and Compliance with Pollution Control Norms

Abstract Through a review of various published reports from the Central Pollution Control Board, Comptroller and Auditor General of India, the Ministry of Environment and Forests and the Infrastructure Development Finance Company, the chapter presents the water quality status of selected stretches of some important rivers and their tributaries; analyses the changes in water quality status over a time (2007–2014); and examines the status of compliance of pollution control norms by the statutory agencies.

Keywords Pollution control norms · Biological oxygen demand · Faecal coliform · Aquifers · Aquatic resources

3.1 An Overview of Water Quality Status in India

Water quality is an important indicator for monitoring the environmental changes which have direct impacts on social and economic development (OECD 2008). Non-availability of fresh water resources and contamination of available water are major problems that hinder socio-economic development.

Wastewater and effluents from industrial processes and commercial establishments, as well as from domestic sources, are contaminating lakes and rivers. About 66 million people spread over 22 states, particularly in rural areas, are facing water quality risks due to high fluoride levels, and around 10 million high arsenic levels, in their groundwater, due to overdraft. Groundwater also gets contaminated by salt, iron, nitrates and other chemicals (Husain et al. 2013, 2014; Nickson et al. 2007; Srikanth 2009). Pollutants that enter aquifers remain there for years and spread through the water, leading to groundwater over large areas becoming unsuitable for consumption and other human uses.

Understanding the characteristics and mechanisms of contamination of surface and groundwater can help policymakers evaluate the effectiveness of water management measures and create an environment for sustainable development. In its report, 'Status of Water Quality in India', the CPCB (2013a) identifies the locations in rivers, lakes, ponds, tanks and observation wells which do not meet water quality criteria with regard to organic pollution, measured in terms of BOD and coliform bacteria count. According to the CPCB, 63% of the observations indicate a BOD less than 3 mg/L, 19% between 3 and 6 mg/L and only 18% above 6 mg/L. About 50% have total coliforms over 500 MPN/100 ml, and 35% have faecal coliform over 500 MPN/100 ml.

3.2 Compliance with Pollution Control Norms

The Infrastructure Development Finance Company (2011) reported that less than a third of the estimated 38,254 million litres of sewage generated each day in Class I and II towns in India was getting treated. The situation is made worse by the abstraction of water for irrigation, drinking, industrial use and thermal power generation, which leaves very little freshwater in rivers and other water bodies to dilute the pollutants. The 320,000 small-scale industrial units in India contribute almost 40% of the pollution from all industrial sources. Urban areas in India generate more than 100,000 metric tonnes of solid waste every day, part of which also ends up in water bodies, either directly or by leaching. Also causing harm are medical waste and medicines past their expiry dates, which in many cases are not disposed-off safely. Runoff from mines and industrial areas also causes water pollution, as does agricultural runoff, which contains pesticide and fertilizer residues. An indication of the magnitude of the nonpoint source pollution is the nitrate concentration in groundwater, which exceeds permissible limits in Haryana and Punjab.

A report by the Comptroller and Auditor General of India (2011) on the performance of statutory agencies for pollution control in India concluded that the overall planning for the control of pollution on the part of the MoEF and the states falls short of the ideal. In an ideal situation, a complete inventory would be undertaken of rivers and lakes and the keystone species associated with them; pollution levels of rivers and lakes in terms of biological indicators should be measured; contaminants and human activities that impact water quality in rivers, lakes and groundwater should be identified and quantified; the risks of polluted water for health and the environment should be assessed; a basin-level approach to control of pollution should be adopted; and water quality goals and corresponding parameters should be developed and enforced for each river and lake.

The report noted that this shortfall would have repercussions on the implementation and outcomes of programmes for control of pollution. The performance of projects undertaken under the National River Conservation Plan (NRCP) was observed to be unsatisfactory, the National Lake Conservation Plan (NLCP) to be ineffective in conserving and restoring lakes in India, and the inspection and monitoring of projects under the NRCP and NLCP to be inadequate at the local, state and central levels (Comptroller and Auditor General 2011).[1]

In the light of the comprehensive recommendations of the report, in May 2011 the MoEF constituted a committee to consider these recommendations and prepare a roadmap for their implementation. The committee's proposals included a time-bound action plan to address capacity issues related to sewage treatment, an amendment to the Environment (Protection) Act of 1986 to link penalties for contravention of the act, strengthening of the Water Quality Assessment Authority and constitution of a State-Level Monitoring Committee. Seven years later, untreated faecal sludge and septage from cities is still the single biggest source of water resource pollution in India (Ministry of Urban Development 2017b).

The Comptroller and Auditor General (2011) notes, 'River cleaning and control of pollution programmes for our polluted rivers are being implemented since 1985. The programmes seek to address pollution from point and non-point sources through construction of Sewage Treatment Plants, low cost sanitation, electric crematoria etc. However, the data on the results of these programmes are not very encouraging.' It also says that 'most lakes in India are under threat from nutrient overloading which is causing their eutrophication and their eventual choking up from the weeds proliferating in the nutrient-rich water. Implementation of NLCP in conserving these lakes has had no discernible effect.' With respect to groundwater, 17 states had identified pollution levels in terms of arsenic, nitrate, salinity, and so on (Comptroller and Auditor General 2011). In 2013, both the NRCP and the NLCP was merged into a new scheme called National Plan for Conservation of Aquatic Ecosystems with the idea of holistic conservation and restoration of lakes and wetlands for achieving desired water quality enhancement (Press Information Bureau 2013).

The norm for DO is 6 mg/L or more, and that for BOD (5-day BOD at 20 °C) is 3 mg/L or less, for drinking water sources without conventional treatment but after disinfection. Faecal coliform (MPN/100 mL) should be less than 500 for treated water disposed-off in water or land and in class B water fit for outdoor bathing (https://cpcb. nic.in/water-quality-criteria). The water quality in rivers upstream and downstream of selected cities over a period of time was analyzed. BOD and faecal coliform are mostly well above, and DO well below, the norms, especially downstream of cities. Importantly, there was no improvement in these values between 2007 and 2013, in spite of the many programmes for the rejuvenation of rivers. Faecal coliform in Ganga River downstream of Varanasi is a staggering 100,000, though the norm is 500 (CPCB 2016). If this is the situation of the Ganga, when there is a national mission for cleaning the river, what can be said of the other rivers in the country?

A similar trend can be noticed regarding lakes as no basin-wise planning is undertaken. As for the projects undertaken under the NLCP to restore and conserve lakes, after their completion, the water quality had improved in only a few lakes, including

---

[1]Central assistance to the tune of 70% was provided to build new wastewater treatment systems and upgrade the technology and capacity of existing ones; the other 30% was to come from the state government.

Nainital Lake in Uttarakhand, Kotekere and Sharanabasaveshwara Lakes in Karnataka and Mansagar Lake in Rajasthan. The project was suspended due to the inability of authorities to stop the very large volume of sewage into the lake, which made it unresponsive to bio-remediation. The tables in the Appendix (Table A.1, Table A.2, Table A.3, Table A.4, Table A.5 and Table A. 6) give the water quality of selected stretches of some important rivers/tributaries in India.

3.3   Findings and Conclusions

In India, the Central and State Pollution Control Boards are responsible for monitoring aquatic resources (rivers, lakes and groundwater) in India for pollution control, and are also responsible for setting effluent standards and enforcing pollution control norms. However, they do not have the power to penalize polluters. Faced with the dual task of monitoring water quality and enforcing pollution control norms, they have not been very effective as regulatory bodies. They maintain a large network of water quality monitoring stations (for select physical, biochemical and bacteriological parameters), covering rivers, groundwater and lakes. But lack of adequate technical manpower for routine functions—collecting and testing water samples, and interpreting the results—further reduces their effectiveness.

The Ministry of Urban Development evaluates cities on the basis of cleanliness, sanitation and municipal documentation (Swachh Survekshan). While freedom from open defecation and the availability of toilets are among the evaluation criteria, management of the wastewater from toilets, including sewage collection and treatment and the presence of sewage networks, is not considered.

Discharge of sewage into surface water bodies, which is widely prevalent in India, is a major threat to freshwater availability and water security; about 62% of the total sewage generated in India is discharged directly into nearby water bodies without treatment.

The BOD, faecal coliform and DO of river water downstream of the cities of Hyderabad, Delhi, Agra, Mathura and Varanasi do not conform to norms. And there was no improvement in these values from 2007 to 2013, in spite of the many programmes for the rejuvenation of rivers (CPCB 2013b). This is true even for the Ganga, though there is a national mission specifically to clean up this river.

Chapter 4
Effectiveness of Wastewater Collection and Treatment Systems

Abstract This chapter discusses the situation in Indian cities and towns with regard to wastewater generation, collection and treatment systems. The statistics pertaining to water supply by urban water utilities, wastewater generation, sewage collection, existing capacities for treatment of wastewater and actual volume of sewage treated in the existing facilities are presented and analyzed across city classes, i.e., metropolitan cities and class I and class II cities. While comparing the wastewater treatment capacity available across states, issues pertaining to low capacity utilization of wastewater treatment plants are also discussed.

Keywords Water supply · Wastewater generation · Sewage collection · Wastewater treatment capacity utilization

4.1 Wastewater Generation and Collection in India's Urban Areas

As per the Census of India (2011), 31.6% of the population (380 million people) live in urban areas. A town or urban agglomeration (UA) with a population of 100,000 or above is categorized as Class I, while Class II towns/UAs have between 50,000 and 99,999. Around 70% of India's urban population live in Class I towns/UAs. There are 468 Class I towns/UAs, 53 of which have a population of one million or above and house 42.6% of the urban population. Greater Mumbai UA (18.4 million), Delhi UA (16.3 million) and Kolkata UA (14.1 million) have more than 10 million and are known as the mega-cities. As the population of India continues to rise, the proportion of people living in urban areas is also rising. It is expected to grow by 404 million by 2050, to around 780 million (United Nations 2014). While the population growth in the mega-cities has slowed considerably in the past decade, the population is expected to grow faster in smaller towns and cities and the newly created towns.

On average, 80% of the freshwater supplied to a city or town gets converted into wastewater. There is wide variation between states and between cities in per capita availability of piped water supply and sewerage networks. A better sewerage network ensures that all the wastewater generated is collected, increasing the quantum of

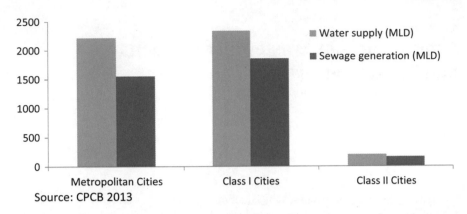

Source: CPCB 2013

Fig. 4.1 Water supply and wastewater generation in different classes of cities

wastewater as a proportion of the water supplied. Figure 4.1 shows average water supply and wastewater generation per city in different city categories. A smaller proportion of the water supplied is collected by the sewerage system in metropolitan cities[1] (around 70%) than in Class I and II cities (79.4% and 80%, respectively). This might be because in some metros, like Mumbai, many of the people living in slums do not have sewerage connections, though they can access water from common public sources.

4.2   Sewage Treatment Capacity

In 2015, the sewage generated by cities and towns in the country was estimated at 61,754 million litres per day (MLD), as against the developed sewage treatment capacity of 22,963 MLD (Central Pollution Control Board 2016). Hence, the average treatment capacity was 37%. The technologies used by most sewage treatment plants (STPs) are primary settling, followed by Activated Sludge Process; 2) Up flow Anaerobic Sludge Blanket + Polishing Pond; or a series of Waste Stabilization Ponds (WSPs). There is a big gap between the volume of wastewater generated and the treatment capacity available for domestic wastewater and for industrial wastewater, which mainly generated by small-scale industrial units. Figure 4.2 gives an indication of the wastewater treatment scenario in different classes of cities in India. The extent of treatment of collected sewage is much higher in metros (around 71%) than in Class I and II cities. It is least in Class II cities, where only 13% of the collected wastewater is treated.

In 2005, the CPCB reported that the treatment capacity for domestic wastewater was 18.6% of sewage generated, and only 72.2% the STPs treatment capacity was being utilized, and therefore only 13.5% of the sewage generated was being treated.

---

[1]Cities having population more than 1.0 million.

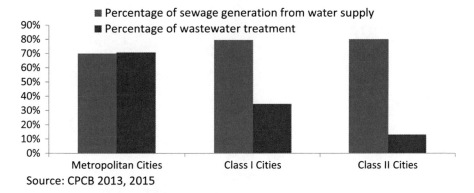

Source: CPCB 2013, 2015

Fig. 4.2   Water supply and wastewater treatment scenarios in different classes of cities

These statistics clearly indicated dismal position of sewage treatment, which is the main cause of pollution of rivers and lakes. Operation and maintenance of existing plants and sewage pumping stations is also a very neglected field, as nearly 39% plants are not conforming to the general standards prescribed under the Environmental (Protection) Rules for discharge into streams (Central Pollution Control Board 2005). In 2009 the treatment capacity was reported to be 11,800 MLD, making it 30% of the sewage generated. But capacity utilization was only 65%, mainly because of the lack of proper networks for transport of sewage to the treatment plants. This meant that 20% of the total quantity of 38,000 MLD of sewage generated in metros and in Class I and II cities/towns was getting treated, showing a bit of an improvement for the four years. By 2015, treatment capacity had increased substantially, to 22,963 MLD. But this was partly offset by the increased rate of urbanization, and thus a major jump in the volume of sewage, with the result that treatment capacity as a percentage of sewage generated has not increased much. CPCB (2016) highlighted that because of the hiatus [*sic*] in sewage treatment capacity, about 38,791 MLD of untreated sewage (62% of the total sewage) is discharged directly into nearby water bodies', partly due to the lack of ways for the sewage to reach the treatment plants. Thus, there was not much improvement in the status of wastewater treatment in the country in 2015, compared to 2005.

Figure 4.3 shows the total amount of sewage generated from cities (metros and Class I and II cities) and the total STP capacity in different Indian states (based on CPCB 2013b). The total sewage treatment capacity was estimated at around 18,000 MLD, against a total volume of 49,750 MLD of sewage collected. While Maharashtra has the largest volume of sewage generated (13,953 MLD) and the largest sewage treatment capacity (6897 MLD) in these three categories of cities, sewage treatment capacity as a percentage of sewage collected is highest in Gujarat (69.3%), followed by Delhi (61.3%). According to the CPCB (2013b), the STPs in Maharashtra could treat only 49% of the sewage collected, even if all of it is brought to the STPs. The state of Karnataka has a sewage treatment capacity of only 2%, Uttar Pradesh, which

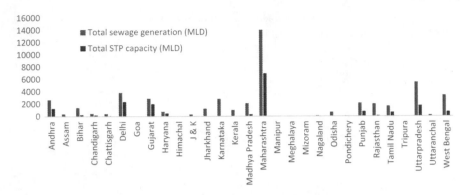

Fig. 4.3 Total sewage generation and STP capacity in Indian states

is in the Ganga basin, has only 31%, and West Bengal, only 21%. However, data on plant capacity utilization are not available for all these plants.

The MoEF had funded 179 STPs under the Ganga Action Plan I, Yamuna Action Plan I and NRCD schemes. The total installed capacity of STPs under the NRCD schemes was 4864.6 MLD. Inspection of 152 STPs (with a total capacity of 4716 MLD) was carried out, and the actual utilized capacity of these STPs was assessed as 3126.42 MLD, which is 66% of the total installed capacity. Figure 4.4 shows the total capacity of the plants and the amount of wastewater treated, by state (data from CPCB 2013b). In Gujarat and Punjab, the capacity utilization is quite high, while in Maharashtra, Tamil Nadu, Uttar Pradesh and Haryana it is quite low.

To elaborate, all the metropolitan cities have a sewerage system, but only a third of the Class I cities and less than one-fifth of the smaller urban centres do. In all urban centres, only part of the population is covered by the sewerage system. As of 2005, the system in some cities did not function properly or was defunct; other cities had a sewerage system but did not have STPs to treat the wastewater, resulting

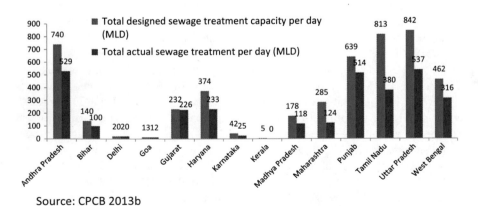

Source: CPCB 2013b

Fig. 4.4 Statewise sewage treatment capacity and volume of wastewater treated, by state

in untreated sewage being discharged into water bodies (Central Public Health and Environmental Engineering Organisation 2005). In 2012, as many as 987 out of every 1000 households in India had no access to a toilet with a piped sewer system (National Sample Survey Office 2014). Considering that rural areas (which do not have sewerage networks) accounted for only 62% of the total population of the country at that time, this figure meant that the sewage generated by a significant proportion of the population in urban areas never got collected.

4.3 Sanitation Access in Urban India

The extent of use of modern sanitation systems greatly depends much on access to water. Access to piped water in urban areas varies widely across states, from 20% of the population in Bihar, 30.2% in Assam and 34.9% in Kerala, to 96.8% in Chandigarh and 95.5% in Himachal Pradesh. But there is not so much variation in the percentage with access to water for all household purposes. Access to water for use in toilets also does not show such wide variation across states as most of the toilets either do not have a water connection or a reliable water supply. Of the urban households in Bihar, Assam and Kerala, 82.6%, 91.1% and 99.2%, respectively, have access to water for use in toilets from other sources.

Along with modern improved toilets, building a sewer network is extremely important for environmental sanitation. The part of the urban population with sanitary toilets varies from 77.7% for Jharkhand to 100% for Uttarakhand and Arunachal Pradesh, with a national average around 89%. But the number of administrative wards in the urban areas connected to the sewer network shows a much wider variation, from nil in Meghalaya to 92.1% in Delhi and 93.2% in Gujarat (Ministry of Statistics and Programme Implementation 2016). The actual proportion of toilets connected to sewer networks could be lower. Wherever toilets are not connected to a sewer network, various on-site sanitation systems are in use, mainly septic tanks. With more people migrating to cities and towns, leading to higher population density in these urban centres, on-site sanitation systems can pose greater health risks, with high risk of contamination of groundwater.

But the Swachhta Status Report (Ministry of Statistics and Programme Implementation 2016) provides data on disposal of household wastewater, but not waste from toilets (black water). The census of 2011, which the report was based on, collected information on the system of disposal of wastewater, including kitchen wastewater and bath and wash water, but not the wastewater from latrines. The Swachh Survekshan (Ministry of Urban Development 2017) evaluated cities on cleanliness, sanitation and municipal documentation. While freedom from open defecation and the availability of toilets were among the criteria used for assessment, management of wastewater from the toilets was not mentioned. There was also no mention of access to the sewerage network or of wastewater treatment.

4.4   Findings and Conclusion

Contrary to general perception, analysis of secondary data shows that a smaller proportion of the water supplied (around 70%) is collected by the sewerage system in metropolitan cities as compared to Class I (79.4%) and Class II cities (80%). Of the wastewater collected, in metros 71% is treated, compared to 35% in Class I cities and 13% in Class II cities (CPCB 2013b). It can be inferred that the infrastructure available for wastewater treatment is quite poor in Class I and II cities, compared to metros.

Since 2005, the government of India has taken concerted action to prevent water pollution. As a result, the capacity of wastewater treatment systems in the cities and towns of India was significantly enhanced between 2005 and 2015, through this was partially offset by the increase in effluent generation. In 2005, the treatment capacity for domestic wastewater was 18.6% of sewage generated, with actual capacity utilization of 72.2%. By 2009, treatment capacity was 11,800 MLD, or 30% of the sewage generated—but capacity utilization was only 65%. By 2015, treatment capacity had increased to 22,963 MLD; but as of 2016, only 38% of the collected sewage was getting treated (CPCB 2016).

There are three major centralized wastewater treatment systems in use in India: ASP, UASB, and WSP. ASP is the most common in large metros and Class I cities; WSP is the most common in Class II cities. ASP has the highest efficiency (99%) in BOD and COD removal, followed by facultative/maturation pond (96%) and sludge blanket reactor (90%).

Out of the total of 17,848 MLD of treatment capacity reported in 2013 from the three categories of cities across India, Maharashtra had the highest capacity, at 6896.7 MLD. Karnataka had one of the lowest, at 55.6 MLD. But capacity as a proportion of total sewage generation was highest in Gujarat (69.3%), followed by (60.3%), and in Maharashtra it was only 49% (CPCB 2013b).

Data on the capacity utilization available for 152 STPs (funded by the MoEF) in metropolitan cities and Class I and II cities show an overall capacity utilization of about 65%. The main reasons for this are problems with O&M and the absence of sewage networks. There is significant variation in capacity utilization across states, with Gujarat recording the highest (97%) and Maharashtra the lowest (43.5%). Despite the projects undertaken under the NLCP to restore and conserve lakes, in only a very few lakes has the water quality improved. Treated wastewater is used for horticultural purposes and by industrial units in Chennai, New Delhi, Bengaluru and Hyderabad. However, the water thus used is only a very small fraction of the treated wastewater that is being generated. A major constraint is in the transport of treated water to localities where demand exists.

Chapter 5
Health Impacts of Water Pollution and Contamination

Abstract This chapter lists the range of water-borne diseases, causative organisms, mode of spread of the disease-causing agents/organisms and the symptoms. It also presents the data on reported cases of water-borne discusses in India, the major diseases that cause death and loss of productivity. It finally presents case studies from several pollution hotspots in India, showing the impact of source water contamination on public health, and the agents causing pollution of water bodies in those localities.

Keywords Water-borne diseases · Causative organisms · Symptoms · Bacteria · Virus · Protozoa

5.1 Overview of Public Health Impacts of Water Contamination in India

Poor-quality water spreads disease and hampers socio-economic progress. *Escherichia coli*, Shigella and *Vibrio cholerae* bacteria, Hepatitis A, poliovirus and rotavirus, and parasites like *Entamoeba histolytica*, Giardia and hookworm are the pathogenic organisms mainly responsible for water-borne diseases in India (WaterAid 2016b). Untreated sewage can be the source of large numbers of pathogenic microorganisms such as bacteria, viruses and protozoa parasites (US EPA 2003), which are causative agents of diseases (Al-Rekabi et al. 2007; Wani et al. 2013).

Some 195,813 habitations[1] in the country are affected by poor quality water for human consumption (CPCB 1999). Water-related diseases are an economic burden on both households and the nation. At the household level, the economic loss includes the cost of treatment, while the government spends a lot of money and time on treating the sick and providing other supportive services. On an average, rural people in India spend at least Rs. 100 a year on the treatment of water-and-sanitation-related diseases (WaterAid 2016b). According to the government of India, this adds up to Rs. 67

---

[1] As per the office of the Registrar General and Census Commissioner, Government of India, Village or Town is recognised as the basic area of habitation.

billion annually, which is just Rs. 520 million less than the annual budget of the Ministry of Health and Family Affairs and more than the allocation for education (WaterAid 2016b).

The diseases and health conditions which result in mortality include diarrhoea, typhoid, cholera, paratyphoid fever, dysentery, jaundice, amoebiasis, malaria and dengue. According to one estimate, India loses 73 million working days every year to water-borne diseases (Tewari and Bapat 2016). According to a report on the cases and deaths by water-borne diseases in India, submitted to the Lok Sabha, the country registered 69.14 million cases of four water-borne diseases in the five years from 2013 to 2017 (Tripathi 2018). Diarrhoea caused 6514 deaths, the most of water-borne diseases in India, in the same period. Other killers were viral hepatitis (2143), typhoid (2061) and cholera (20).

A list of bacterial, viral and protozoan diseases, with causative organism, mode of spread and symptoms, is given in Table 5.1. Similar details for water-related vector-borne diseases are given in Table 5.2.

Chemicals present in the water can have negative effects on health. Pesticides washed into rivers and drinking water bodies and systems can have adverse effects on the nervous, reproductive and endocrinal systems. Phosphates, organophosphates and carbamates can cause cancer. Blue baby syndrome is common among babies drinking milk when there are nitrates in the water. Lead affects the central nervous system. Arsenic causes liver damage and skin cancer. Fluorides yellow teeth and damage spinal cords. Petrochemicals, even in smaller quantities, cause cancer. These are the end results of water pollution. Many groundwater and surface water sources are polluted with heavy metals, persistent organic pollutants, and plant nutrients, harming people's health. Fresh drinking water sources are not only threatened by over-exploitation and poor management but also by ecological degradation.

Table 5.3 provides data on the incidence of various water-borne diseases reported in India and the deaths they caused from 2013 to 2017.

5.2   Case Studies on Water Pollution and Impacts on Public Health

This section presents some case studies on water pollution and its effects on human health.

1. *Drinking Water Contamination in Ludhiana, Punjab*

Bedi et al. (2015) reviewed water quality problems and examined the incidence of various water-related diseases and their economic impact on households in Ludhiana, Punjab. As per the information available on the basis of discussions with the officials of the Punjab Water Supply and Sewerage Board, Bedi et al. (2015) estimated that about 80% (which is 39,955) of the water related diseases (including diarrhoea, cholera, malaria, jaundice, skin infection, etc.) in Ludhiana district between January

Table 5.1 Water-borne diseases: bacterial, viral and protozoan

Bacterial diseases			
Disease	Causative organism	Mode of spread	Symptoms
Bacterial			
Typhoid	*Salmonella typhi*	Contaminated food, water, milk, unwashed raw vegetables, flies	Continuous fever which increases day by day; temperature higher in evening than morning; body ache, headache and constipation; haemorrhage from ulceration in small intestine
Cholera	*Vibrio cholerae*	Water or food contaminated by bacteria from stool of cholera patient	Painless diarrhoea, vomiting, 30–40 stools per day, which soon become watery and colourless, with flakes of mucus floating in them
Bacterial diarrhoea	*Shigella* spp.	Contaminated food, water, direct personal contact	Diarrhoea, with blood and dysentery mucus in the stool, along with severe gripping pain in the abdomen; stools not too frequent (4–10 per day), faecal matter scanty; patient looks ill
Leptospirosis	*Leptospira*	Rodents, primary hosts, carry organisms in kidneys; infection by wading or swimming in water contaminated with rodent urine	Fever, pain in legs, nausea, vomiting; congestion of the conjunctival blood vessels around the corneas
Viral			
Infectious hepatitis	Hepatitis A	Food and water contaminated with virus in stool	Loss of appetite, nausea, vomiting and diarrhoea, fever; urine dark-coloured; eyes and skin appear yellow

(continued)

Table 5.1 (continued)

Bacterial diseases

Disease	Causative organism	Mode of spread	Symptoms
Protozoan			
Amoebic dysentery	*Entamoeba histolytica*	Ingestion of cysts in food and water	Abdominal discomfort and diarrhoea, with or without blood or mucous in stools, fever, chills and gripping pain in abdomen
Diarrhoea	*Giardia intestinalis or Giardia lamblia*	Food or water contaminated with faeces having cysts	Intestinal disorders leading to epigastric pain, abdominal discomfort, loss of appetite, headache, loose bowels
Bilharzia (schistosomiasis)	*Schistosoma* spp.	Cercaria larvae of flukes in water penetrate skin of persons wading in water	Allergy-like itch, rash, aches, fever, eosinophilia; when infection becomes heavy, eggs may block arterioles of lungs causing schistosomiasis and possible congestive heart failure
Guinea worm	*Dracunculus medinensis*	Unfiltered water	Blister near the ankle, causing allergy and aches

Source https://www.iasabhiyan.com/water-borne-diseases/

2011 and May 2012 were caused by unsafe water. Out of this, about 12,000 cases of water-related disease were registered in a six-month period (January–May 2012) itself.

Further, analysis by Bedi et al. (2015) suggest that the number of people affected by water-related diseases was highest in slums, followed by LIG (low income group) and MIG (middle income group) dwellers. In 2011, in the sampled households 18,258 instances of disease from drinking unsafe water were recorded. Barring slums, the mixed and low-income groups adopted coping mechanisms like installing Reverse Osmosis machines and Aqua Guards (filtration machine for removal of microbes from water). The VHG (very high-income group), HIG (high income group), MIG and LIG households spent Rs. 18,400, Rs. 11,375, Rs. 9667 and Rs. 8500, respectively. However, the percentage of sampled households adopting such mechanisms is quite

Table 5.2 Water-related diseases: vector-borne

Disease	Causative organism	Vector	Hosts	Symptoms
Malaria	Plasmodium spp.	Female Anopheles (primary or final host)	Man (intermediate host)	Shivering, chills and sweating; as chills subside body temperature rises as high as 106 °F; when temperature comes down patient sweats profusely and becomes comfortable, until next attack; they take place at regular intervals
Filaria (elephantiasis)	Wuchereria (Filarioidea)	*Culex fatigans*	Man (final host)	Enlargement of limbs and scrotum
Dengue	Parvovirus	*Aedes aegypti*	Man (reservoir)	Sudden onset of moderately high fever; excruciating joint pain; intense pain behind eyes; a second rise in temp following brief remission; reduction in neutrophilic white blood cells

Source https://www.iasabhiyan.com/water-borne-diseases/

low. The investment cost per household was Rs. 8000, Rs. 7667, Rs. 6167 and Rs. 4625, respectively, for those four groups. The annual costs for treating water-borne diseases per household was Rs. 540, Rs. 25,540, Rs. 15,178, Rs. 111,376, Rs. 23,681 and Rs. 6073, respectively, for VGH, HIG, MIG, LIG, mixed, and slums (Bedi et al. 2015).

The study concluded that drinking water, though pure at the source, is polluted during its transmission. Despite various interventions by government institutions,

Table 5.3 Reported cases (in numbers) of water-borne diseases and deaths in India, 2013–2017

Disease	2013		2014		2015		2016		2017	
	Cases	Deaths	Cases	Deaths	Cases	Deaths	Cases	Deaths	Cases	Deaths
Cholera	1130	5	844	5	913	4	718	3	385	3
Acute diarrhoea	11,414,000	1629	11,748,631	1137	12,913,606	1353	14,166,574	1555	9,230,572	840
Typhoid	1,650,145	387	1,736,687	425	1,937,413	452	2,215,805	511	1,493,050	286
Viral hepatitis	110,125	574	138,554	400	140,861	435	145,970	451	98,086	283

Source Tripathi (2018)

water-purifying mechanisms are needed by households. The economic cost of consuming contaminated water leading to additional monetary burden, the water-related diseases still persist conspicuously.

2. *Pollution in Allahabad City, Uttar Pradesh*

Allahabad is faced with human health hazards due to water pollution and environmental degradation. The under-developed pockets of the city suffer due to lack of safe drinking water supply. The diseases arriving in the villages through polluted water and improper disposal of excreta add to the problem of sanitation. As the city of Allahabad is in transition, but except in the core area, the city is lacking urban and rural planning, and drainage and sanitation systems. Rapid industrialization is causing further stress on the environment.

A study assessed water pollution and its impact on human health in Allahabad City (Pandey et al. 2017). Empirical observation of the behaviour of locals towards the polluted environment was undertaken before generating data for analysis. The main types of surface water pollution were suspended solids, organic matter, biological decomposition of organic matter, and inorganic dissolved salts. The river water at Allahabad was unfit for drinking or bathing, with BOD exceeding 6.4 mg/L. According to the city development plan, in 2015 the city's sewage output was 250 MLD. Only 60 MLD was treated, and the rest flowed into the Ganga and Yamuna.

The main source of groundwater pollution were heavy metals, total dissolved solids, fluorides and arsenic from industry, and nitrates and pesticides from agriculture. Most of the sites in Allahabad had chloride above the permissible level in water (250 mg/L), and pH ranged from 6.9 to 8.2. Factors identified as causes of deteriorating groundwater quality were effluent from industrial units and solid waste. Both give rise to toxic elements which penetrate deep into the ground, contaminating groundwater. At the time of the study, the city had only 22% sewage coverage (for 44,300 households). There is no sewage treatment for the rest. Most of the toilet flushes are not connected to soak pits.

3. *Leather Tanning*

The leather tanning industry needs an aqueous medium, and a tannery's effluent contains the discharge from drums, pits or paddles containing various soluble and insoluble materials. In conventional tanning, for every 10 tonnes of hide and skin processed, 2–3 tonnes of salt is discharged while pickling. Untreated tanning effluent released into streams causes pollution, which also seeps into the ground, degrading groundwater in an 8 km radius.

The daily production of wastewater from the 3000 tanneries in India (from large-scale to cottage units) is approximately 175,000 m^3 (Kodandaram et al. 1972).

The pollutants cause release of gasses into the environment, creating a nuisance, and in some cases the intense odours cause nausea among people. In streams, the effects are severe, including high salinity and harm to aquatic life. Groundwater is degraded and rendered unfit for drinking. The resulting solids, in the form of salt dust, when washed away by rainwater, may also pollute the groundwater.

The impact of tannery wastewater on water bodies has been extensively studied by Central Leather Research Institute and National Environmental Engineering Research Institute and many other organizations in India (for instance refer Mondal et al. 2005; Roy 2012). This industry is marked as a 'red category industry[2]' for its pollution effect (Roy 2012). The Ganges in the north, the Cauvery in south and the wetlands of Kolkata are polluted by the tanneries of Uttar Pradesh, Tamil Nadu and West Bengal, respectively.

Recently, some tanneries report adopting new technologies for 'water-free' tanning. But we are a long way from all tanneries doing so. Recently the government of Uttar Pradesh ordered all tanneries closed for three months during the Kumbh Mela religious festival.

5.3 Findings and Conclusions

As review suggests, pollution of source water, both groundwater and surface water, contamination of water during transmission, and poor environmental sanitation cause public health hazards through water-borne diseases, including typhoid, cholera and leptospirosis (from bacterial infection); infective hepatitis (from virus); and diarrhoea, amoebic dysentery, Guinea worm and bilharzia (from protozoa).

India loses 73 million working days annually to water-borne diseases. The country registered 69.14 million cases of four water-borne diseases in the five years from 2013 to 2017. Diarrhoea, the biggest killer, caused 6514 deaths during this period (Tripathi 2018). The case studies of pollution hotspots in Allahabad, Ludhiana, and tanneries in Uttar Pradesh, Tamil Nadu and West Bengal confirm the high prevalence of water-borne diseases caused by water pollution and environmental degradation (WaterAid 2016b).

---

[2]Industries having a pollution index score of 60 and above are called red category industries. The pollution index score reflects the level of pollution that can be caused by a particular type of industry. The red category industries are critically polluting industries.

Chapter 6
Reuse of Treated Wastewater: Present Scenario

Abstract This chapter presents some reported cases of reuse of treated wastewater from urban areas of India for agriculture in peri urban areas, watering of public parks, and municipal uses, including the extent of use. It then discusses some of benefits of using treated wastewater. Finally, it highlights some infrastructure and policy related issues that keep the demand for treated wastewater in India suppressed.

Keywords Reuse · Agriculture · Industry · Recycling · Peri-urban areas

6.1 Introduction

The major reuse of treated wastewater is in urban areas for gardening and watering of trees, and irrigation in peri-urban areas. Reuse of treated wastewater in agriculture instead of direct use of untreated effluent also helps prevent soil degradation, groundwater contamination, and human health hazards associated with handling of contaminated water. Many city administrations have started moving in this direction.

As agriculture does not requires high quality water with regard to physical, chemical and bacteriological concentrations, there is growing recognition in India of the need to treat wastewater from urban areas for reuse in agriculture, to reduce the pressure on the available freshwater resources, for which there are competing claims from the domestic, industrial and power sectors. In this chapter, we will discuss some of the well documented cases of wastewater reuse.

6.2 Cases of Reuse of Treated Wastewater

The Haryana Urban Development Authority plans to reuse treated wastewater on a large scale. It proposes to use treated wastewater for public parks and green belts, as well as activities like construction. In fact, a portion of the treated water from the Dhanwapur plant is bought by builders for construction activities, and some of the water that is allowed to flow into the Yamuna is used by farmers for irrigation (Pant 2018).

The New Delhi Municipal Council is constructing decentralized STPs of different capacities at 10 locations in the municipal area, with plans to use the treated water for horticulture in the neighbourhoods. The council has also identified 12 schools where decentralized STPs of varying sizes, depending on the available area, will be set up. Delhi aimed to treat and reuse 25% of the total sewage produced by 2017, increasing it to 50% by 2022 and to 80% by 2027 (PwC 2006). Though recent studies show that around 12,000 farmers in Delhi reuse treated wastewater irrigating an area of nearly 1,200 ha (Amerasinghe et al. 2013), no data on the volumetric extent is available. The National Capital Region Planning Board (NCRPB), in its Regional Plan 2021 for National Capital Region (NCRPB 2005), says that all new development areas should have two distribution lines, one for drinking water and other for non-drinking water (recycled wastewater treated for reuse), and that all the water requirements for non-drinking purposes in big hotels, industrial units, central air conditioning of large buildings and institutions, large installations, irrigation of parks and green areas, and other non-potable demands should be met through treated recycled wastewater.

The Chennai Metropolitan Water and Sewerage Board (CMWSB) sells secondary-treated sewage to industrial units, and the Chennai Municipal Corporation uses secondary-treated sewage to water plants and lawns in public parks and traffic islands. Part of the treated effluent has been auctioned off and used in the cultivation of para grass (for use as fodder) since the formation of the CMWSB in 1978. The STPs commissioned in 2005 and 2006 are self-supporting, without any need for power from the electricity grid (Urban Design Research Institute n.d.). The present treatment capacity of the STPs under the CMWSB is 486 MLD (Elangovan 2010). This is sufficient to treat the wastewater generated from the water supplied by the board, the actual water use in the city is much higher because it includes supply from private sources, resulting in untreated sewage entering the stormwater drains and water bodies.

The Hyderabad Metropolitan Water Supply and Sewerage Board offers treated domestic wastewater from its STPs for non-potable uses, as in construction. The board charges Rs. 125 per 5000 litres, with the tanker and transportation charges to be borne by consumers (*Telangana Today* 2017). All the treated effluent was earlier diverted into the Musi River, which farmers have been lifting for irrigation purposes.

In Bengaluru, the Bangalore Water Supply and Sewerage Board supplies 40 MLD of tertiary-treated water to the Kempegowda International Airport. The water is sold at Rs. 25 per kilolitre. Surat, Nagpur and Vishakhapatnam are other cities where treated wastewater is used by industrial units (Sahasranaman and Ganguly 2018).

As observed in Hyderabad, one major factor influencing the viability of this model is the transferability of treated wastewater. The water supply board or municipal corporation does not provide any facilities to move treated water to the farmers' fields, so farmers have to do it. In such situations, only those farmers close to a treatment plant or those who require a small volume of water for high-value uses (fruit trees and plantation crops) will find it viable to transport water by tanker to their fields. Still, there are lawns and plantations maintained by the city councils (municipalities and corporations), industrial and corporate houses and commercial establishments, including hotels, which require water throughout the year. These are the situations which would normally demand treated wastewater.

6.3 Concluding Remarks

Currently, though large amount of treated wastewater is available even from cities falling in water-scarce regions, not all is consumed in the neighbouring region due to insufficient demand. Lack of adequate infrastructure for transport of treated water to the areas where it is required is a factor limiting the demand. Lack of awareness among the farmers about the long-term benefits of using treated wastewater over untreated wastewater is another constraint. Absence of government policies that encourage reuse of wastewater is also an issue. The fundamental issue is that there are no economic incentives for industries and agriculture and commercial establishments to prefer treated wastewater over water from conventional sources. The recent studies point towards these (source: based Amerasinghe et al. 2013; Elangovan 2010; NCRPB 2005; Pant 2018; PwC 2006).

Given the fact that treated wastewater is emerging as an important component of the overall water balance, the state governments need to frame policies on wastewater reuse to derive the maximum None of the Indian states had a policy on reuse of wastewater—whether in the domestic sector, agriculture, industry or commercial use. As noted by Kumar (2018a), in future, the volume of treated wastewater available from city peripheries is going to be large, with large-scale water imports, increasing pressure on urban water utilities to comply with environmental management safeguards, and greater willingness on the part of urban dwellers to pay for environmental management services as a result of rising per capita incomes (Kumar 2018a). Another recent paper estimates that if all the urban water demands are met from the available supplies, the total amount of wastewater generated from the cities/towns by 2040 will be 7.87 million hectare-metre (78.7 BCM) (Kumar 2018b).

Social and economic benefits from reuse of treated water, and to create financially viable models for investment in wastewater treatment.

Chapter 7
Wastewater Treatment Technologies and Costs

Abstract The chapter presents data on different wastewater treatment technologies that are in use in the urban areas of India, including centralized and decentralized ones, and the extent of use of each technology. It also presents data pertaining to the performance of different technologies, especially their BOD and COD removal efficiency, and discusses the comparative performance. The capital and operation and maintenance costs of various treatment technologies are also presented and discussed.

Keywords Activated sludge process · Waste stabilization pond · Anaerobic pond · Septic tank · Conventional wastewater treatment · Decentralized wastewater treatment · BOD removal efficiency · COD removal efficiency

7.1 The Context

Centralized treatment of wastewater is normally used in urban areas where the water supply is usually adequate to generate sufficient wastewater for its flow in the sewer system. However, in many cities, the areas that are not covered by the public water supply and sewerage networks, decentralized wastewater treatment systems are in use. As discussed in Chapter 4, the treatment capacity for domestic wastewater is currently inadequate in the country. There is also a gap in the treatment of industrial wastewater, which is mainly generated by small-scale industrial units. Wastewater treatment system can be classified as conventional (centralized) and decentralized. Open defecation is highly prevalent in rural areas in India. Given the dispersed and unplanned development in rural areas and the absence of the technical expertise that would be required to operate them, it is impractical to provide centralized sewage treatment systems in rural areas. So, various types of decentralized systems are in use there, the most important being pit latrines and septic tanks.

In this chapter, we will present: the statistics on different types of conventional and decentralized wastewater treatment systems that are in use in Indian cities; and available analysis of their technical and economic performance. The key findings emerging from the comparative analysis of the performance of different types of treatment systems are also presented.

7.2   Conventional or Centralized Wastewater Treatment

Various technologies can be used to treat wastewater. But only a few of the STPs in India are functioning satisfactorily. In urban areas, many are non-functional due to various problems associated with the centralized treatment system. Thus, due to lack of operational and maintenance facilities, untreated and semi-treated wastewater flows into rivers, causing severe health and environmental problems. STPs are also used for decentralized wastewater treatment in urban areas for small townships, apartment complexes and even individual apartment blocks.

Centralized wastewater treatment has high operation and maintenance costs. Experienced technical personnel are required to implement and operate the system. It requires significant electrical power, and does not work where continuous electricity is not available. On average, the operation and maintenance cost for treatment of sewage is estimated at Rs. 30,000 per month per MLD, with advanced technologies being even more expensive. Activated sludge treatment requires 2.6 kW of electricity per MLD of sewage (CPCB 2013b).

'Conventional' wastewater treatment separates solids from liquids by physical processes and then purifies the liquid using biological and chemical processes. The process has three phases (mechanical, biological and chemical), which are referred to as primary, secondary and tertiary treatment. The purpose of primary treatment is to separate solids from liquids as much as possible, producing a homogeneous liquid that can be treated biologically, and a sludge that can be disposed of or treated separately.

Primary treatment removes large objects and reduces oils, grease, sand, grit and coarse solids. This is usually done using large sedimentation tanks and rotating screens to remove floating and larger materials.

Secondary treatment is intended to degrade organic compounds that consume oxygen when degraded and therefore increase the BOD and COD of the water. To do this, most treatment plants in developed countries use the activated sludge process, in which the liquid is heavily oxygenated and a substrate is provided so that naturally occurring bacteria and protozoans consume the biodegradable soluble organic compounds. These microorganisms also bind less soluble fractions into floc particles, which tend to settle to the bottom of the tanks. Eventually, the microorganisms also flocculate and settle, so that the supernatant liquid can be discharged.

Tertiary treatment is the final stage to raise the effluent quality to the standard required before it is discharged. This phase usually includes various types of filtration, total removal of nutrients and chemical disinfection.

7.3   Treatment Technologies Used in Sewage Treatment Plants: Performance and Costs

In Class I cities, the activated sludge process (ASP) is the most commonly employed technology, with 59.5% of total installed capacity, followed by upflow anaerobic sludge blanket (UASB), with 26%—these two technologies being mostly used as the main treatment unit of a scheme including other primary or tertiary treatment units. A series of waste stabilization ponds (WSPs) are employed in 28% of the plants, though their combined capacity is only 5.6%. ASP technology is the most suitable for large cities because it requires less space than UASB and WSP, both of which are land-intensive (CPCB 2013b).

In Class II towns, a series of WSPs is the most commonly employed technology, with 71.9% of total installed capacity and 72.4% of STPs, followed by UASB, with 10.6% of total installed capacity and 10.3% of STPs (CPCB 2013b).

The Biological Oxygen Demand (BOD) and Chemical Oxygen Demand (COD) removal efficiencies of STPs vary according to the climate of the region. They also depend on how the STP is operated and maintained. The ASP, facultative/maturation pond and SBR (sequential batch reactor) technologies have the highest BOD removal efficiency, at 99%, 96% and 90% respectively. The BOD removal efficiency of other technologies like oxidation pond (OP), sequential batch reactor, trickling filters, UASB and WSP varies between 60 and 75%. Anaerobic lagoons and stabilization ponds have the lowest, at 35%. ASP and facultative/maturation pond have the highest COD removal efficiency (92% and 88%, respectively); other technologies vary from 45 to 65%. Sequential Batch Reactor (SBR) COD removal efficiency is very low, at 46% (CPCB 2013b). Thus, SBR is efficient for the removal of biological matter but not inorganic matter from wastewater. Anaerobic lagoons and stabilization ponds are inefficient for removal of both biological and inorganic matter. Figures 7.1 and 7.2 compare the performance of various technologies in terms of BOD and COD removal, respectively.

Treatment technologies adopted under NRCD-funded schemes include the natural system, conventional technology and advanced technology. Table 7.1 summarizes the treatment technologies adopted in the NRCD schemes (CPCB 2013b).

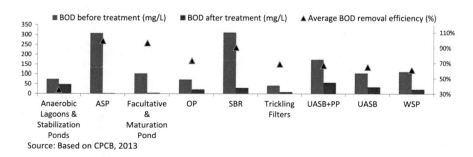

Source: Based on CPCB, 2013

Fig. 7.1 Performance of different wastewater treatment technologies in BOD removal

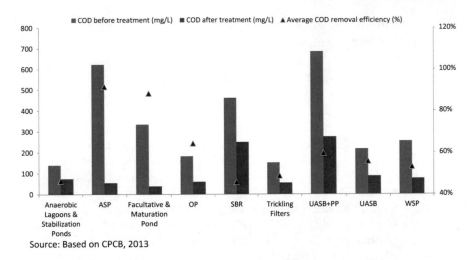

Source: Based on CPCB, 2013

Fig. 7.2 Performance of different wastewater treatment technologies in COD removal

Table 7.1 Conventional technologies used in sewage treatment plants in India (number of plants)		
Oxidation pond (OP)	34
Waste stabilization pond (WSP)	31
Anaerobic lagoons	3
Activated sludge process (ASP)	19
Extended aeration	2
Trickling filters	7
Cyclic activated sludge process	2
Upflow anaerobic sludge blanket (UASB)	37
Karnal technology	4
Sequential batch reactor (SBR)	3
Fluidized aerobic bioreactor	2
Biochar	2
Others	6

Source CPCB (2013b)

   A comparison of the cost and area required for the various technologies used in sewage treatment is given in Table 7.2.

Table 7.2 Cost comparison of various technologies used in sewage treatment plants

Technology	Average capital cost (secondary treatment), Million Rs./MLD	Average capital cost (tertiary treatment), Lac Rs./MLD	Total capital cost (secondary + tertiary), million Rs./MLD	Total area (m^2) per MLD secondary + tertiary treatment	Average total daily power requirement, kWh/d/MLD	Total annual O&M costs, Lac Rs./MLD, up to secondary treatment
Activated sludge process	6.80	40	10.8	1000	185.7	353.02
Moving bed biological reactor	6.80	40	10.8	550	223.7	372.11
Sequential batch reactor	7.50	40	11.5	550	153.7	288.15
Upflow anaerobic sludge blanket + extended aeration	6.80	40	10.8	1100	125.7	290.72
Membrane bioreactor	30.0		30.0	450	302.5	–
Waste stabilization ponds	2.30	40	6.30	6100	5.7	116.09

Source CPCB (2013b)

7.4 Decentralized Systems for Wastewater Treatment: Performance and Costs

Decentralized wastewater treatment systems can be built to suit site conditions and varying inflows. The sewer networks required are shorter, and a variety of simple and natural treatment methods can be used which need no mechanization. As a result, it is possible to achieve cost-effective sanitation, in addition to reusing the wastewater and nutrients near the source of waste generation.

The houses in rural areas are not built in a planned manner and are not close together like in urban areas. Houses may be in the middle of large farms, so that they are isolated. Hence, in rural areas decentralized wastewater treatment systems are more suitable. Currently, even in urban areas of India about 70% of the population is served by on-site sanitation systems. The advantages of decentralized wastewater treatment systems are low operation and maintenance costs, zero or very low energy consumption depending on the technology chosen, easy operation by semi-skilled persons, and economic returns depending on the technology used. There is safe reuse of treated effluent as well as of sludge. A decentralized system is suitable for all site conditions, including an undulating topography where conventional sewage treatment cannot be implemented.

The performance of the various treatment systems depends on the influent characteristics and temperature. The performance can be defined by the approximate BOD removal rate: 25–50% for septic and Imhoff tanks; 70–90% for anaerobic filters and baffled septic tanks; and 70–95% for constructed wetland and pond systems (Kumar 2014).

Decentralized systems for wastewater treatment can be divided into three broad categories:

- Soil-based systems, which include subsurface infiltration, rapid infiltration/soil aquifer treatment, overland flow, and slow-rate systems;
- Aquatic systems, which include waste stabilization ponds and floating aquatic plant systems; and
- Wetland systems, which include free water surface and subsurface flow systems.

The septic tank is the most favoured type of on-site sanitation system in India; pit latrines are also used, largely in rural areas. A septic tank is a watertight, multi-chambered receptacle that receives black and/or grey water and separates the liquid from the solid waste, which it stores and partially digests. (Black water is wastewater from toilets, and grey water is wastewater from bathing, kitchen and other household activities, excluding toilets.) A septic tank is a combined sedimentation and digestion tank. The settleable solids in the sewage settle at the bottom in one or two days, accompanied by anaerobic digestion of settled solids (sludge) and liquid, resulting in reasonable reduction in the volume of sludge, reduction in biodegradable organic matter, and release of gases like carbon dioxide, methane and hydrogen sulphide. The effluent, although clarified to a large extent, will still contain appreciable amounts of dissolved and suspended putrescible organic solids and pathogens, as the efficiency is only 30–50% for BOD and 60–70% for total suspended solids (TSS) removal. This effluent can contaminate groundwater. Septic tanks also need to be de-sludged for effective functioning.

Most cities and towns in India do not have effective septage or faecal sludge management mechanisms. Faecal sludge is the settled contents of pit latrines and septic tanks, while septage is the liquid and solid material that is pumped from a septic tank, cesspool, or on-site treatment facility after it has accumulated over a period of time. For a pit latrine or septic tank system to be environmentally safe there must be a proper system for faecal sludge management. Table 7.3 lists the treatment methods and costs involved in selected emerging decentralized wastewater treatment technologies where the treated water can be used for horticultural or other purposes.

Table 7.3 Cost comparison of different decentralized sewage treatment technologies

Method	Description	Capital cost	O&M costs	Remarks
Bioremediation	Using biological products to decompose organic matter	Rs. 20,000–30,000/MLD for flowing water, Rs. 400–5000/ML for still water	Rs. 1.9 Lac/MLD for flowing water, Rs. 2.8 Lac/acre for still water	Suitable for in situ treatment of lakes and ponds
Decentralized wastewater treatment systems (DEWATS)	Sedimentation, anaerobic treatment, plant root zone treatment, oxidation process	Rs. 35,000–70,000/kLD	Rs. 1000–2000/kLD per year	Selection of modules is based on the quality of wastewater to be treated; no electro-mechanical equipment used
Fixed film biofilter	Settling and flow equalization followed by biochemical process	Rs. 25,000–30,000/kLD	1000–2000/kLD per year	Use of microculture reduces the area required due to enhanced natural degradation of contaminants
Phytorid	Settling followed by plant root zone treatment involving both aerobic and anaerobic treatment in specially engineered baffle treatment cells	14,000–35,000/kLD	Rs. 100–2000/kLD per year	Locally available wetland plants are used

Source Centre for Science and Environment (2014)

7.5 Findings and Conclusion

Both centralized and decentralized wastewater treatment systems are in use in Indian cities and towns. As regards the performance of centralized systems, the Biological Oxygen Demand (BOD) and Chemical Oxygen Demand (COD) removal efficiencies vary according to the climate of the region. They also depend on how the STP is operated and maintained. The ASP, facultative/maturation pond and sequential batch reactor technologies have the highest BOD removal efficiency, at 99%, 96% and 90% respectively. The BOD removal efficiency of other technologies like oxidation pond (OP), sequential batch reactor, trickling filters, UASB and WSP varies between 60 and 75%. Anaerobic lagoons and stabilization ponds have the lowest, at 35%. ASP and facultative/maturation pond have the highest COD removal efficiency

(92% and 88%, respectively); other technologies vary from 45 to 65%. Sequential Batch Reactor (SBR) COD removal efficiency is very low, at 46% (CPCB 2013b).

As regards capital cost of centralized systems, of the six treatment technologies considered for analysis, membrane bioreactor is the costliest, at Rs. 30 million/MLD capacity, followed by SBR, at Rs. 7.5 million/MLD. Regarding O&M cost, among the five technologies for which data are available, moving bed biological reactor is the most expensive, at Rs. 37.2 million/MLD capacity, followed by ASP, at Rs. 35.3 million/MLD. For WSP, both capital and O&M costs are the lowest, at Rs. 2.3 million/MDL and Rs. 11.6 million/MLD, respectively (CPCB 2013b).

The septic tank is the most favoured on-site (decentralized) sanitation system in India. They help reduce the volume of sludge, biodegradable organic matter, and release of gases like carbon dioxide, methane and hydrogen sulphide. The effluent, however, will still contain appreciable amounts of dissolved and suspended putrescible organic solids and pathogens, as the efficiency is only 30–50% for BOD and 60–70% for total suspended solids (TSS) removal.

The decentralized wastewater treatment systems are bioremediation, DEWATS, fixed film biofilter and Phytorid. Their capital cost ranges from Rs. 0.4–5.0 million/MLD for bioremediation; Rs. 35–70 million/MLD for DEWATs; Rs. 25–30 million/MLD for fixed film biofilter and Rs. 14–35 million/MLD for Phytorid. Hence, they are costlier than centralized systems, in spite of being less efficient. Their operation and maintenance costs are much lower than that of centralized plants, ranging from 0.28 million/MLD for bioremediation; Rs. 1–2 million/MLD for DEWATs and fixed film biofilters; and Rs. 0.1–2.0 million for Phytorid.

Chapter 8
Case Studies on Performance of Wastewater Treatment Systems

Abstract The chapter presents case studies on the performance of sewage treatment plants from four locations in India, viz., New Delhi, Pune, Nashik and Hyderabad. The various performance attributes, including technical (BOD and COD removal, bacteriological concentration of treated water, dissolved oxygen and pH), managerial (staff, water quality monitoring) and economic (capital and O&M costs per cubic metre of water treated) aspects are analyzed and discussed. The factors influencing plant operating efficiency are also briefly discussed.

Keywords Sewage treatment plant · BOD removal efficiency · COD removal efficiency · Bacteriological concentration · Plant operating efficiency

8.1 Introduction

In order to get an in-depth understanding of the characteristics of the wastewater treatment systems, their operation and maintenance, their technical performance and the range of physical, technical and institutional factors influencing the overall performance, a detailed survey was carried out in selected locations. For this, visits were made to the plant sites for physical verification, and key officials who supervise the plant operations were interviewed on details of plant functioning, including the plant capacity, investments, operating efficiency, manpower requirements to run the plant, operation and maintenance (O&M) costs, technical and institutional issues with efficient operation of the plant, water quality and overall plant performance monitoring, and so on. The survey covered plants in four cities: New Delhi, Hyderabad, Pune and Nashik. Of these, New Delhi, Hyderabad and Pune are metropolitan cities, with a large population (more than a million) and many treatment plants.

8.2 General Profile of the Plants and Locations

A general profile of the plants covered in the survey is provided in Table 8.1. As per the analysis of the primary data collected and discussions with the STP operators, the degree of treatment of wastewater is quite high, with 80–95% reduction in the

Table 8.1 Features and key performance indicators of the wastewater treatment plants

Location	Plant name	Plant features		Cost of the plant (crore rupees)	Cost of rehabilitation/ upgrading (crore rupees)	Year started/ upgraded	Capacity (MLD)	Average loading (MLD)	Degree of BOD/COD removal (%)	pH of treated wastewater	Dissolved oxygen in raw/treated water (mg/L)	Bacteriological concentration in treated water (count/100 mL)
		Type of treatment	Name of treatment system									
Pune	Dr. Naidu Hospital Plant	Aerobic	Activated sludge process	40.15	2010	DNA	115.00	95–100	96/93	7.1	Test not done	Test not done
Hyderabad	Amberpet	Anaerobic and aerobic	UASB reactor + aerated lagoons, polishing pond	DNA	95.00	Pre-independence/ 2008	339.00	280	90/90	7.7	0.29/4.04	2,251
Nashik	Ramtekdi STP Plant	Anaerobic	UASB reactor	14.40	14.00	2003/2010	130.00	105–110	80-82/80	7.6	DNA/4.5	110
Rohini, New Delhi	Rithala Phase II Plant	Aeration and filtration	High-rate aeration with biofiltration	81.93	12.74	2002/2011	181.84	185	85/86.4	7.3		Test not done

DNA: data not available

Source Based on the analysis of the primary data and information collected form the STP operators in 2019

BOD and COD of the effluent. The pH is within the permissible range for disposal in natural water bodies. In two cases for which data are available, the DO content also meets the standards for survival of aquatic life. In Nashik, the treated wastewater is discharged into the Godavari River, which is then used by the thermal power station at Eklahare, near Nashik City. The farmers downstream also use the flowing water for irrigation.

In Pune, all the treated wastewater is discharged into the Mulla-Mutha River, and the treated wastewater is reused for irrigation by farmers and also by the Pune Municipal Corporation for watering trees, with the help of tankers. In Hyderabad, the extent of reuse of the treated wastewater is quite low, with only some commercial establishments buying the water. Transportation of water to the locations of demand is an issue in the city, as the plant does not have any conveyance facility. The rest of the water is discharged into the Musi River.

In Nashik, all the treated wastewater is reused. Though there was a plan to sell the treated wastewater to a private agency, the Water Resources Department of Maharashtra objected, claiming that the Nashik Municipal Corporation could not sell the water to a third party and should release the treated wastewater into the river. So, there is no trading of treated wastewater. In Delhi, the treated wastewater is supplied to Pragati Power Corporation Ltd. for their plant at Bawana, to North Delhi Power Ltd. for their owner plant at Rohini, to Delhi Development Authority for horticulture. The extent of reuse is about 28 MLD (as of 2016). The treated wastewater can be used for agricultural and industrial purposes and also to water nearby parks and gardens, and a market can be developed for this.

In all four cases, the plant operation is outsourced to private companies by the municipal corporation or board. In Nashik, each year a competitive bidding process is followed for outsourcing the O&M work. In Pune, the operator is VA Tech Wabag Ltd, Kolathur Chennai. In Hyderabad, it is Ramky & Co. Ltd. In Delhi, it is Suez India Private Ltd. In Nashik, it is Mahajan Brothers Nashik which runs the plant.

8.3 Working of the Wastewater Treatment Plants

Operation and Maintenance

The details of plant operation, including staff employed for operation, labour engagement, and annual O&M costs for the individual plants are presented in Table 8.2. The O&M costs are normally estimated per unit volume of treated wastewater. The total annual O&M cost for Pune plant is Rs. 20.94 million. For the Nashik plant, it is Rs. 13.6 million. Based on the total volume of water treated in the plant per day, the unit O&M cost of treating wastewater works out to be Rs. $0.57/m^3$ for the Pune plant and Rs. $0.338/m^3$ for the Nashik plant. This does not include the unit capital cost, which cannot calculate as we do not have a proper idea of the life of the system. For the two other plants, data on O&M are not available.

Table 8.2 Details of plant operation, number of staff engaged, annual operation and maintenance charges, and electricity charges

Plant location	Type of treatment	Details of plant operation (human resources)				Is plant operation outsourced?	Total no. of people employed by the agency	Annual operation and maintenance details (Rs.)			
		Officer in charge	If there a plant supervisor?	No. of plant operators	No. of labourers employed			Chemicals (for plant and lab)	Salaries	Repair and maintenance	Electricity charges
Pune	Aerobic	Executive engineer of PMC	Yes	12	35	Yes	53	557,040	1,37,83,967	27,81,923	3,823,230
Hyderabad	Anaerobic and aerobic	General manager	Yes	30	65	Yes	115				
Nashik	Anaerobic		Yes (DE, JE)	3	35	Yes	42	112,000	9,388,867	500,000	3,600,000
New Delhi	Aeration and filtration	Executive engineer of Delhi Jal Board	Yes (plant manager)	DNA	DNA	Yes	7[a]				Self-generated

DNA: data not available

[a]The staff employed by the private agency are: plant manager, safety officer, maintenance manager, electrical engineer, process in charge, mechanical engineer, and biogas engineer/service engineer

Source Based on the analysis of the primary data and information collected form the STP operators in 2019

Water Quality Monitoring

All the four plants have well-equipped water quality testing laboratories, where tests are performed for up to (in one case) 14 parameters. In Pune, tests are carried out for BOD, COD, TSS, pH and MLSS at the lab maintained by the Pune Municipal Corporation. In Hyderabad, the Hyderabad Metropolitan Water Supply and Sewerage Board is in charge of the technical supervision and administration of the STP, whereas the other responsibilities are outsourced. Tests for 14 water quality parameters are carried out by the board daily. The lab has two water quality analysts (chemist/microbiologist) with MSc degrees, a lab assistant, a lab attendant, a data entry operator, and a field assistant. A third-party performance evaluation is carried out monthly by an analyst from the Telangana SPCB.

In Delhi, the quality of treated wastewater is tested once in a day in the lab of the Delhi Jal Board; four parameters are analyzed (BOD, COD, TSS and pH). In Nashik, testing is done once a week for total dissolved solids, pH, suspended solids, chloride, sulphate and detergent. There is a monthly evaluation of plant performance by the Maharashtra Pollution Control Board.

Factors Influencing Plant Operating Efficiency

In Pune, problems are encountered with the sewage collection system. There are some areas which are still unconnected, especially new villages that have been included in the jurisdiction of Pune Municipal Corporation. In some places, there is leakage from the main line. Old or inadequate transmission lines cause sewage to flow in open *nullahs* (natural drainage channels). Funding is needed for closed sewage lines. Similar problems are encountered in Hyderabad, where the full plant capacity is not utilized. In the Delhi plant, electricity is generated from the biogas produced in the plant, and all the electricity required to run the plant is met from the bio gasifiers. Hence the O&M cost is drastically reduced.

As per the discussions with the STP operator, in Nashik City, also, there are problems with sewage collection. Some areas are still unconnected, especially new villages that have been included in the jurisdiction of the municipal corporation. In some places, there is leakage from the main line. As of 2019, the corporation has a total capacity of 342 MLD for treating wastewater, and nine STPs are functioning. About 1,400 km of sewerage line has been laid to collect wastewater from the city. The corporation is constructing two more STPs with a total capacity of 68 MLD. These STPs are expected to be commissioned before 2020.

Upcoming Investments in Wastewater Treatment

Funds for investment in wastewater treatment plants and associated infrastructure do not appear to be a problem in any of the locations. For instance, in Pune, a loan agreement was signed in 2016 by the government of India and the Japan International Cooperation Agency for pollution abatement of in the Mula-Mutha River in Pune by January, 2022.[1] Under the agreement, the government of Japan committed to

---

[1] The Mula-Mutha is one of the 302 polluted river stretches of the country identified by the CPCB. The major reasons for pollution of Mula-Mutha are discharge of untreated domestic wastewater into

providing India a soft loan of 19.064 billion yen (about Rs. 1,000 crore) for the project at 0.3% annual interest.

The major components proposed under the project include construction of 11 new STPs (which will add 396 MLD of treatment capacity to the existing 477 MLD), laying of 113.6 km of sewer lines and renovation/rehabilitation of four existing intermediate pumping stations. The final STP capacity of 873 MLD will be enough to handle the sewage generation projected for the year 2027 (Pandey 2016).

8.4 Findings and Conclusion

Detailed analysis of the primary data collected from the four wastewater treatment plants with secondary treatment facilities (one each from Delhi, Pune, Hyderabad and Nashik) and with capacity ranging from 115 MLD to 390 MLD was carried out. The plants have adequate capacity to handle the sewage being collected at the site; and they are adequately staffed; but their capacity is not fully utilized due to inadequate infrastructure to collect and transport sewage. The technical performance of these plants *vis-à-vis* the level of treatment of effluent is quite high—from 80 to 95% reduction in BOD and COD. In most cases, the treated water is reused mainly in agriculture.

As per the information obtained from the surveyed STPs, the unit O&M cost of treating wastewater works out to Rs. $0.57/m^3$ for the Pune plant and Rs. $0.338/m^3$ for the Nashik plant. In all cases, the municipal corporation or board which owns the plant outsources the work of plant operations to private agencies, but has water quality testing laboratories under its supervision to test the effluent quality. Testing involves from four to 14 parameters. The State Pollution Control Boards of the respective states test water quality monthly to assess the performance of these plants.

---

the river due to inadequate sewerage system (including pumping stations) and sewage treatment capacity in the town, as well as open defecation on the river banks.

Chapter 9
Environmental Sustainability and Economic Viability

Abstract This chapter discusses the factors influencing the economic viability of wastewater treatment plants. The range of physical/environmental and socio-economic factors that determine the cost of treatment per unit volume of wastewater are first discussed. Subsequently, the range of physical and socio-economic factors that determine the economic value of the treated wastewater which includes that accrued from direct economic, environmental and social benefits. The trade-off between economic viability and environmental sustainability of wastewater treatment is also highlighted.

Keywords Cost of wastewater treatment · Economic value of treated wastewater · Climate · Geohydrology · Temperature · Land prices

9.1 Introduction

The building of water infrastructure in India's urban areas is not keeping pace with the rapid population and economic growth. Availability of wastewater treatment infrastructure is highest in the metropolitan cities, but even there the gap between the volume of wastewater generated and the volume treated is considerable. On the other hand, existing capacity is not fully utilized, especially in Tamil Nadu, Maharashtra and Uttar Pradesh. In most cases this is due to lack of pipelines to bring sewage to the treatment plants. Of the 152 STPs across various cities in India, 34 are non-operational. Often, not enough money is set aside for O&M costs, even when considerable amounts are spent on new infrastructure. Lack of adequate staff for O&M and water quality monitoring (including water sample testing and lab maintenance) are also issues. This leads to poor maintenance of STPs, and to their discharge not conforming to environmental protection standards (CPCB 2013b).

While there is a major emphasis on the construction of toilets under the Swachh Bharat Abhiyan, not enough attention is paid to the treatment of the wastewater produced or to faecal sludge management. Apartment complexes for high-income people are beginning to have their own STPs, but this is too expensive to be implemented in all housing developments. Users must also be aware of the interrelationship between hygiene and health and on how their actions can lead to pollution of land and water bodies.

New wastewater infrastructure needs to be built, and existing wastewater treatment systems improved. Upgrading of the wastewater treatment technologies also needs to be undertaken, for better treatment of the sewage. There is a need for periodic renovation of the sewer system, and for extending the sewerage network. New wastewater treatment systems need to be constructed with more focus on economic viability and environmental sustainability. Importance should also be given to raising awareness of sanitation and pollution issues among users, to improve cooperation in maintaining their own environments.

9.2 Environmental Sustainability Versus Economic Viability

Wastewater treatment systems should be environmentally sustainable. One important factor hindering environmental management performance of wastewater treatment systems is economic viability. Higher levels of treatment of wastewater to improve the environmental performance would increase the cost of the plant, but may not increase the value of the benefits proportionately, especially the direct economic benefits, thereby adversely affecting the economic viability of the plant.

The economics of wastewater treatment is determined by the cost of the treatment system and the economic value of the treated water. For most land-based wastewater treatment systems, such as WSP, ASP, anaerobic ponds, soil aquifer treatment and oxidation ditches, the cost is a function of the land area required for treatment. In many of India's cities, land prices are rising to astronomical levels. Here the economically viability of the system improves if the amount of land required for treatment is less, but the technology is capable of providing the required level of treatment. But the land area required for treatment and therefore the cost is dependent on: the technology used for treatment, which is decided by the physical factors such as temperature, altitude, soils and geo-hydrology. The land area required, and therefore the cost of removal per unit weight of pollutant, also depends on the degree of treatment required, which is a function of the pollutant concentration in the influent. In certain situations, where land prices are high, even membrane technology, which provides the highest level of treatment (tertiary treatment), might be cost-effective.

Table 9.1 illustrates the second point, about the effect of degree of treatment on the unit cost of pollution reduction, using data from effluent treatment plants of 53 sugar firms. For the same volume of influent load, the marginal cost of removal of 100 g of BOD to bring the effluent to a concentration of 30 mg/L is a function of the degree of reduction in BOD required (Pandey 1997). This means, for higher BOD concentration of the influent, the cost of treatment per 100 gm of BOD is higher.

Table 9.1   Cost of pollution abatement in terms of BOD removal for different BOD concentrations (Rs./KL)

Wastewater flow (KL)	Marginal cost of treatment of wastewater to the minimal national standards (MINAS) level of 30 mg/L		
	Minimum BOD concentration	Average BOD concentration (992.45 mg/L)	Maximum BOD concentration (1,200 mg/L)
Minimum (500)	1.22	1.55	1.69
Average (1,335.8)	0.93	1.18	1.30
Maximum (2,500)	0.79	1.00	1.09

Source Pandey 1997

Assessing the economic value of the treated water is not easy. It depends on the kinds of opportunities that exist for reuse of wastewater (Hanjra et al. 2015). Four scenarios can be observed in India: (1) very high demand for the wastewater for irrigation, where currently untreated wastewater is used; (2) no demand for wastewater for irrigation or fisheries, so it has to be disposed into streams; (3) no demand for wastewater in irrigation, but some water demand for growing fish; and (4) untreated wastewater is not used for irrigation or fisheries, due to high toxicity.

The first situation is most common in arid and semi-arid areas of the country. There is considerable use of wastewater in the peripheral areas of big cities (such as Delhi, Hyderabad, Ahmedabad, Bangalore, Vadodara and Jaipur) for agricultural production, mostly leafy and other vegetables for the urban market. The second situation is possible in humid, high-rainfall regions like Kerala and Assam, where plenty of freshwater is available in surface water bodies for irrigation and fisheries, and irrigation water demand is low. The third situation is common in sub-humid areas like Kolkatta and Bhubaneswar, in eastern India. Here, communities use wastewater for irrigation not because freshwater bodies are not available, but because the wastewater contains nutrients from food waste. The fourth situation is encountered in industrial areas, where the trade effluent has toxicity high enough to cause soil degradation, and where farmers have alternative sources of irrigation water, as we found in Tirupur, in the Noyal River basin of Tamil Nadu.

In the first case, the cost of treating the wastewater to the desired level should be less than the incremental benefits of (a) reducing health risks to the irrigators; (b) reducing ecological damage caused by degradation of the soil; and (c) reducing groundwater quality deterioration. In the second and third case, the cost of the treatment system should be less than the benefits of improving the general environmental quality of the stream/river and aquatic life (positive ecological and economic externalities). In the fourth case, the cost of the treatment system should be less than the marginal economic outputs of crops irrigated with wastewater and the reduction in public health risks associated with unsafe disposal of effluents into the streams, which is the positive externality associated with improvement in environmental conditions.

It has been well demonstrated around the world that there are economic and environmental benefits from reuse of treated wastewater in agriculture under scarcity conditions, through conservation of freshwater used for irrigation (Rosenqvist et al.

1997); higher agricultural productivity through use of nutrient-laden treated wastewater (Mohammad and Ayadi 2005; Kiziloglu et al. 2007); and less discharge of nutrients into surface water, which in turn reduces the need for costly treatment systems for potability (Rosenqvist et al. 1997), producing positive social impacts.

Murray and Ray (2010) developed a hybrid performance-assessment/decision-support model for the optimal reuse of wastewater for irrigation in Pixian, China, for two irrigation systems, the Xuyan and Zouma, which are facing scarcity of water for agriculture. The model used two scenarios: wastewater reuse to supplement existing irrigation, and wastewater to replace existing irrigation. The coupled performance-assessment and optimization model revealed trade-offs and benefits associated with agricultural reuse. Using wastewater effluent to supplement irrigation could add more than US$20 million in profits for farmers in the Xuyan and Zouma irrigation systems, and using wastewater effluent to replace existing irrigation could conserve over 35 MCM of surface water in local rivers every year. The study found that the added profits would more than cover the cost of building a wastewater treatment plant, which would be about US$2 million for 10,000 m^3 per day capacity.

9.3   Findings and Conclusion

There is trade-off between economics and environmental performance of wastewater treatment systems. High levels of treatment of wastewater to raise the environmental impacts of wastewater treatment systems, which is expensive, may not produce proportional economic benefits.

There are complex considerations involved in the economic evaluation of wastewater treatment systems, that is, in determining both the cost of the systems and the economic value of the benefits of the use of treated wastewater. For land-based treatment systems, the cost depends on the climate, the price of land, and the concentration of the effluent and the degree of treatment required; while the benefits depend on the overall improvement in environmental conditions, willingness to pay for healthy ecosystem, and the economic value of direct benefits derived from the use of the treated water in agriculture, fisheries, and so on.

Chapter 10
Growth of Treatment Plants and Reuse of Treated Wastewater

Abstract In this chapter, we analyze the scale of investment in different types of wastewater treatment systems vis-à-vis the treatment capacity and investment costs, based on projections of the key drivers, viz., urban population growth, presence of water stress, the city category, and demand for treated wastewater. Three different time horizons were considered, i.e., short term; medium term; and long term. The potential extent of reuse of treated wastewater in different types of regions are also analyzed.

Keywords Population growth · Water scarce regions · Water rich regions · Class I cities · Class II cities · Recycling · Reuse · Investment

10.1 Introduction

Over the last 15 years, not only the treatment capacity, but the extent of treatment of wastewater (against the total volume of wastewater generated) has also increased substantially in India. The future growth of wastewater treatment infrastructure and reuse in India will be driven by three main factors. First: the water situation in the cities and towns concerned (whether the city experiences water stress). Second: the size of the city, which also determines the financial capability of the city, overall water demand and the amount of effluent it generates. Third: the type of wastewater treatment infrastructure. All these factors are time dependent. In this chapter, we would systematically analyze these factors to arrive at some estimates of the additional wastewater treatment capacity that would be available in different types of Indian cities at different time horizons, i.e., short term, medium term and long term.

10.2 Future Growth of Treatment Plants and Wastewater Reuse

Based on past trends, we expect that cities and towns in the water-scarce regions will be under enormous pressure to improve the urban environment through wastewater

M. D. Kumar and C. Tortajada, *Assessing Wastewater Management in India*, SpringerBriefs in Water Science and Technology, https://doi.org/10.1007/978-981-15-2396-0_10

treatment and also to reduce the tension arising out of import of water from distant rural areas.[1] Of these, the big cities will complete building wastewater treatment infrastructure in the short term (i.e., in the next one to five years), mostly for secondary treatment of wastewater. The small towns will go for wastewater treatment infrastructure in the medium term (6–15 years), again for secondary treatment.

It is expected that the treated wastewater will mostly be used for agriculture, as the water-scarce regions (mainly western India, north-western India, the South Indian peninsula except Kerala, and the west-central parts of central India) have excessively high demand for water for irrigation. Some will be used to cool thermal power plants.

In the medium term, the big cities will go for recycling of wastewater using tertiary treatment, as the cost of importing freshwater from distant regions will be prohibitive. In the long term (16–30 years), the present small towns are also likely to go for recycling of municipal wastewater, with increasing demand for water and drying up of local sources.

In the water-rich regions, the trajectory will be different for Class I and Class II cities. Since it is easy for them to meet the pollution control norms and environmental management standards at present (with smaller quantities of effluent generated and with large flows in the rivers), adopt treatment technologies only in the medium term, as their populations increase and water quality standards become more stringent. Smaller towns will do this only in the long term, given their poor finances. The treated wastewater will be released into the rivers of the region to maintain their ecological health.

In our assessment, neither the small towns nor the Class I and II cities in water-rich regions will ever go for recycling of wastewater using tertiary treatment, as they are unlikely to face physical shortage of water. But the metros may do so in the long term, as the likely increase in demand for water in other sectors in the next 16–30 years will put enormous pressure on these cities to reduce their dependence on exogenous water. The likely future scenario is depicted in Table 10.1.

To assess the scale of investment that is likely in this sector, we need to understand the geographic distribution of urban population under different city categories. The Census of India (2011) list a total of 7,935 towns in India. Of these, 4,041 were statutory towns (over 5,000 people) and the rest (3,894) were census towns. Of these census towns, 465 were Class I cities, with over 100,000, for an aggregate population of 264.9 million. Of these Class I cities, 53 cities had a population over 1 million (metropolitan cities, or 'metros'). These 53 metros together had 160.7 million people. This was 42.6% of the total urban population in the country (377 million). Some 412 cities (465 minus 53) were Class I cities but not metros. Their total population was 104.2 million. A total of 112.1 million people lived in towns considered Class II or smaller.

---

[1] Already many of the cities in the water-scarce regions of India are importing water from distant regions through inter basin/interregional water transfer projects. A few examples are Ahmedabad, Delhi, Jaipur, Jodhpur, Hyderabad, Chennai and Bangalore. Most often this involves reallocation of water meant for irrigation.

Table 10.1   Projected trajectory of growth in wastewater infrastructure in Indian cities and towns

	Wastewater treatment for reuse			Wastewater treatment for recycling			Remarks
	Short term	Medium term	Long term	Short term	Medium term	Long term	
Water-scarce regions							
Big cities (metro, Class I cities	✓				✓		High growth causing acute shortage of water would force them to go for recycling, and to defer investments in large-scale water transfers, which would be prohibitively expensive
Smaller towns (Class II and below)		✓				✓	
Water-rich regions							
Big cities (metro)	✓					✓	
Class I cities		✓					
Smaller towns (Class II and below)			✓				Smaller towns in water-rich area are unlikely to go for tertiary treatment, as it will not make economic sense

Source Based on authors' own analysis
Note Short term is 1–5 years; medium term, 6–15 years (with a mean of 10 years); long term, 16–30 years (with a mean of 20 years). The base year considered is 2009

Using the Census of India (2011) population data and the information on the average annual renewable water availability in different basins of India (Central Water Commission 2017), we find that 49.8 million people lived in metros in water-rich areas, and the rest (111.1 million) in water-scarce areas. Hence, so far as metros are concerned, about 31% of the urban population live in water-rich regions. We use the same proportion to estimate the number of people living in cities of other sizes. In the water-rich areas, 32.3 million live in Class I cities (excluding metros), and 35 million in Class II and smaller towns. Overall, most of the urban population is likely to be concentrated in the water-scarce regions of the country, which are also economically more prosperous.

We assume per capita wastewater generation of 120 L per day in cities and towns (80% of the average per capita supply of 150 L per day). Using the Census of India (2011) population data and the assumption on daily per capita wastewater generation, we estimate the total capacity requirement for wastewater treatment (for secondary treatment) in the next one to five years at 27,936 MLD (21,960 + 5,976). In the medium term, this is expected to rise by another 13,128 MLD (9,252 + 3,876). In the long term, there will be an additional requirement coming from small towns in the water-rich regions (alone), which we estimate at 4,200 MLD. The demand for tertiary treatment is expected only in the medium and long term, not in the immediate or short term. In the medium term, the demand is expected to come from metros and Class I cities in the water-scarce regions, and the additional capacity requirement is projected to be 21,960 MLD. This would be basically to upgrade the secondary treatment plants or build new (tertiary treatment) plants to replace the existing ones. In the long term, the demand is expected to be generated in small towns in the water-scarce regions (9,252 MLD) and the metros in the water-rich regions (5,976 MLD). The long-term total projected demand therefore is 15,228 MLD (Table 10.2).

However, the populations of these cities and towns will continue to grow in the next few decades, and that will have to be factored in when projecting the additional capacity requirement of the wastewater treatment systems, especially for the investments that are likely to come in the medium and long term. We assume annual compounded growth of 3% for the population of these cities. Hence, for the secondary treatment, the medium-term requirement is revised to 17,643 MLD (considering a mean, 10-year time horizon) and the long-term requirement to 7,586 MLD (considering a 20-year time horizon). For tertiary treatment, the medium-term requirement is revised to 29,512 MLD. The revised, and the long-term requirement to 27,503 MLD (Fig. 10.1).

We estimate the extent of reuse of wastewater in agriculture in the next 10 years at 47,155 MLD, i.e., 29,512 + 17,643 (author's own estimates, Fig. 7), mostly in the water-scarce regions. However, in the medium term, many of these treatment plants will begin to be converted into tertiary treatment systems for recycling of urban sewage, and the treated wastewater will not be available for agriculture and other uses. About 29,512 MLD of water will be reallocated from agricultural use to urban water supply through plant upgrades (refer to Fig. 7, source: authors' own analysis). By 2039, the recycling of wastewater will have increased to 57,015 MLD, with many

Table 10.2 Projected future scale of investments in two types of wastewater treatment systems in different categories of cities

	(Population in million): volume of wastewater for secondary treatment, MLD	Time frame	Volume of wastewater for tertiary treatment (MLD)	Time frame
Water-scarce regions				
Metro and Class I	(183): 21,960	Short term	21,960	Medium term
Small towns (Class II and below)	(77): 9,252	Medium term	9,252	Long term
Water-rich regions				
Metro	(50): 5,976	Short term	5,976	Long term
Class I	(32): 3,876	Medium term	No investment expected	
Class II and below	(35): 4,200	Long term	No investment expected	

Source Based on authors' own analysis
Note Short term is 1–5 years; medium term, 6–15 years; long term,16–30 years

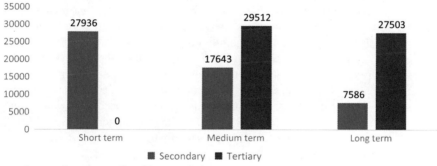

Source: Based on authors' own analysis

Fig. 10.1 Projected additional capacity requirements for wastewater treatment systems (MLD). *Source* Based on authors' own analysis. *Note* 'Short term' considers a time horizon of 1–5 years; for medium term, a time horizon of 6–15 years, with a mean of 10 years is considered and for long term 16–30 years, with a mean of 20 years is considered. The base year considered is 2009

more cities and towns upgrading their treatment plants. Against this, in that year, the amount of treated wastewater available will be 7,586 MLD, and this will be mostly released into rivers in the water-rich regions to maintain their ecological health (refer to Fig. 10.1).

10.3   Future Investments for Wastewater Treatment

In estimating the cost of setting up secondary treatment plants, we assume that capital infrastructure for a secondary treatment facility costs Rs. 3.5 per litre per day of treatment capacity (based on data from primary survey for ASP). Hence, the total capital investment is estimated at Rs. 7,686 crores (US$1.1 billion), Rs. 6,175 cores (US$882 million) and Rs. 2,658 crores (US$380 million) for the short term, medium term and long term, respectively.

The capital cost of a tertiary treatment plant (price of which is close to a membrane-based treatment plant) is assumed to be Rs. 2 million for a capacity of 100 m^3 per day (100,000 L per day) (assuming an average cost of Rs. 20/m^3 for treating wastewater using a membrane-based plant). On that basis, investment for tertiary treatment would be Rs. 59,024 crores (US$8.43 billion) and Rs. 55,006 crores (US$7.85 billion) for the medium term and long term, respectively.

10.4   Findings and Conclusion

There is no single policy at either the central or state level that encompasses pollution prevention, treatment of polluted water for ecological restoration, and reuse of treated wastewater. Also, enforcement of the rules and regulations and the penalties for defiance, non-adherence and violations are not sufficient to ensure compliance.

Our analysis indicates that in the short and medium term, additional wastewater treatment capacity in India is likely to come mainly from cities and towns in the water-scarce regions. In the medium term, large cities in those regions will also upgrade their plants to provide tertiary treatment in response to acute water scarcity. Accordingly, we estimate the total capacity requirement for (secondary) wastewater treatment in the next 1–5 years at 27,936 MLD, in the medium term at 17,643 MLD, and in the long term at 7,586 MLD. For tertiary treatment, the medium-term capacity requirement becomes 29,512 MLD, and the long term, 27,503 MLD.

As per our estimates, the investment in secondary treatment plants is projected to be Rs. 7,686 crores (US$1.1 billion) in the short term, Rs. 6,175 cores (US$882 million) in the medium term, and Rs. 2,658 crores (US$380 million), in the long term. The investment in tertiary treatment plants is projected to be Rs. 59,024 crores (US$8.43 billion) in the medium term and Rs. 55,006 crores (US$7.85 billion) in the long term.

Chapter 11
Market for Treated Wastewater in India

Abstract This chapter analyses the future market demand for treated wastewater, by considering the factors that drive the overall demand (in terms of volume) and the 'willingness to pay' including that for environmental management services in different time horizons. Along with time, the factors considered are: the overall water availability and demand situation in different regions; the suitability of the treated wastewater for various uses; changing income levels of the urban dwellers demand for environmental management services.

Keywords Treated wastewater · Future market potential · Future water supplies · Future water demand · Secondary treatment · Tertiary treatment

11.1 Introduction

The economic viability of wastewater treatment systems would depend on the types of benefits treated wastewater produce which include direct economic, environmental and social (Milieu, WRc and RPA undated), and their value in monetary terms. The environmental and social benefits from treated wastewater that are both direct and indirect cannot always be monetized, unless the communities which derive the benefits pay for the environmental management services (improved quality of river water, improved soil quality, healthy aquatic life, water for recreation, etc.) that it provides, with an influence area whose boundaries are not so clear (Parker and Oates 2016). Economic valuation of these services is problematic, as values (as reflected by the 'willingness to pay') are partly defined by the wealth of the beneficiaries. Wealthier households may be willing to pay more for certain services, as they are able to conceive high value goods more (Korsgaard and Schou 2010).

As regards the direct uses of treated wastewater (say for irrigated agriculture), currently there is willingness to pay for it, and this would increase with the degree of overall scarcity of water, and suitability of the treated wastewater for the particular use. Water scarcity can increase with time. The latter would also change with water scarcity, as acute shortage of freshwater might force water utilities to go for costly treatment options for wastewater to bring it to potable standards which would increase the willingness to pay for that water for domestic uses and municipal services. The overall demand for treated wastewater would also increase with scarcity. In this chapter, we would explore the future demand for treated wastewater in India, with changing water availability-demand situation, and changing income levels and demand for environmental management services.

11.2 Future Market for Treated Wastewater

The demand for secondary-treated wastewater is likely to come from the agriculture sector, which will remain the largest consumer of water, and this demand will be largely driven by the overall scarcity of water. Further, the demand will be greater in regions that are water-scarce and land-rich. In the water-rich regions, there is unlikely to be demand for treated wastewater from the agriculture sector, and the treated water will be released into rivers to meet ecological needs. This is the current trend. The demand for tertiary-treated wastewater will come essentially from the urban water utilities, which will drive the investments in the requisite treatment systems. To assess this demand in volumetric terms, it is important to get a quantitative understanding of the water shortages we are likely to face in future, based on an assessment of water demands and potential supplies.

In recent work which projected the water scenario of India for the year 2040, Kumar (2018b) estimates the total agricultural water requirement at 107.15 million ham by that year. The total water requirement from all sectors under this scenario is estimated at 128.41 million ham. For these estimates, it was assumed that irrigation conveyance efficiency would remain unchanged from the 2015 level (considered by Kumar 2010), that there would be no increase in the yield of the three major crops (wheat, paddy and sugarcane), and that the storage loss of food grain would remain at 15%.

The future water supply potential depends on four main factors: the ultimate water resource utilization potential of the country; the status of development of water resources as of today; the past growth trends in water development; and the poor availability of arable land for irrigation expansion in regions with abundant groundwater, and poor groundwater endowment in high demand growth areas (Kumar 2018b).

The growing concerns about the social and environmental costs of water development projects and the emerging problems associated with water resource exploitation were also considered while extrapolating the past growth trends to arrive at the future levels of exploitation. Kumar (2018b) projected the water supply in the year 2040 as 79.2 million ham. This assumes that surface water utilization potential grows to 82% of the ultimate potential by that year, with annual growth of 0.5 million ham for 22 years. The contribution of canal seepage to groundwater recharge was estimated as 31.85 million ham (under a scenario of low conveyance efficiency). We can make a reasonable assumption that this (seepage) will increase the potential future supplies by an equal amount, i.e., 31.85 million ham. This leaves a gap of 17.36 million ham of water for that year (i.e., 128.41–111.05). This is far more than the total amount of treated wastewater we have projected to be available by that time, 64,601 MLD (2.36 million ham). This entire volume is likely to be used up in agriculture mostly in the water-scarce regions.

While growing environmental concerns will increase the demand for wastewater treatment, their growing economic power will enable large cities or metros (in both water-scarce and water-rich regions) to invest in wastewater treatment technologies in the short term. In water-scarce regions, it will also happen in small towns in the medium term. We have also discussed which regions are likely to witness investments in wastewater treatment systems, the type of treatment plants and the investment trajectory. It is likely that the treated wastewater will end up in the peri-urban areas, producing fruits, vegetables, flowers and forage crops, at a much bigger scale than what is happening today around many cities. Most of the farm produce from these areas ends up in the nearest cities for urban consumers. In Delhi and Kanpur, the municipal corporations are supplying treated wastewater to farmers in peri-urban areas at a fee (Amerasinghe et al. 2013). With greater willingness of farmers in naturally water-scarce regions to pay for treated wastewater to irrigate these high-value crops, financially viable models of wastewater treatment and reuse will emerge.

We have estimated the scale of investment in wastewater treatment (both secondary and tertiary) for different time horizons and the quantum of wastewater that will be treated. Tertiary treatment of wastewater will mainly happen in the metros and Class I cities of water-scarce regions in the medium term, and small towns in the water-scarce regions in the long term, due to growing demand for freshwater supplies from the urban water utilities in those regions and lack of new sources of freshwater. The volume will be 21,960 MLD in the medium term and grow to 57,015 MLD in the long term as smaller towns in these regions also start facing water shortage. In the long term, we also anticipate demand for tertiary-treated water in the water-rich regions, and we project this to be 5,976 MLD.

Regarding secondary-treated water, there will be demand from the peri-urban areas of the metros and Class I and II cities, and even small towns in the water-scarce regions, to meet the growing irrigation demand. Hence, all the treated wastewater, a projected volume of 21,960 MLD, could be purchased by farmers. In the medium term, though the demand for water in agriculture will increase in the water-scarce regions, the actual use will be constrained by the supply, as the cities will recycle the wastewater. We expect total supplies of 9,252 MLD from small towns in the water-scarce regions, so there will be unmet demand for treated wastewater for agriculture. The final scenario *vis-à-vis* water demand by sector and time horizon is presented in Table 11.1.

11.3 Findings and Conclusion

Creating sustainable wastewater treatment systems requires that investments for building them generate adequate benefits that offsets all the costs (Milieu, WRc and RPA undated). Demand for the treated wastewater for various goods (crop irrigation,

Table 11.1 Projected demand for treated wastewater in different categories of cities

	Demand for secondary-treated wastewater (MLD)	Demand sector/time frame	Demand for tertiary-treated wastewater	Demand sector/time frame
Water-scarce regions				
Metro and Class I	21,960	Peri-urban agriculture/short term	21,960	Urban water supply/medium term
Small towns (Class II and below)	9,252	Peri-urban agriculture/medium term	9,252	Urban water supply/long term
Water-rich regions				
Metro	5,976	Ecological functions/short term	5,976	Urban water supply/long term
Class I	3,876	Ecological functions/medium term		
Class II and below	4,200	Ecological functions/long term		

Source Based on Authors' own analysis
Note Short term is 1–5 years; medium term, 6–15 years (with a mean of 10 years); long term, 16–30 years (with a mean of 20 years). The base year considered is 2009

manufacturing processes, watering of lawns and gardens, livestock consumption, etc.) and services that it provides, and the 'willingness to pay' for the same (value attached to these goods and services) are important variables that influences the economic viability. Both are time dependent, and the overall situation with regard to availability and demand for water of adequate quality is a major factor that would drive them. Willingness to pay for the water for various services that produce social and environmental benefits would also be driven by the economic conditions of the communities that benefit.

As per some recent analysis, the projected gap between demand for water in various competitive use sectors and the potential supplies available from conventional sources that include the seepage from irrigation canals, for the year 2040 is 17.36 million ham (173.6 BCM). Further, the gap between water demand and supplies is likely to be high in the water-scarce regions of peninsular, western and north western India (Kumar 2018b).

Therefore, we can expect that the 47,155 MLD (equivalent to an annual volume of 16.7 BCM) of treated wastewater available in the next ten years, to for reuse in agriculture (i.e., in the medium term), mostly in the water-scarce regions. However, in the medium term, many of these treatment plants will begin to be converted into tertiary treatment systems to recycle urban sewage, and the treated wastewater will not be available for agriculture and other uses. Some 29,512 MLD (10.77 BCM) of water will be reallocated from agricultural use to urban water supply through these upgrades. By 2039, recycling of wastewater will have increased to 57,015 MLD (20.81 BCM). Against this, the amount of secondary-treated wastewater available will be 7,586 MLD, and this will be mostly released into rivers in the water-rich regions to maintain their ecological health.

In the current scenario of increasing volume of treated wastewater produced by cities, a policy for wastewater reuse is necessary to derive the maximum social and economic benefits from reuse of treated water, and to create financially viable investment models. Such a policy has to incentivize the 'bulk users' of water in water scarce regions to go for treated wastewater over use of freshwater which has many negative externalities. Incentives can be created if the potential bulk users of water are confronted with a marginal cost for groundwater in the form of a 'resource fee' linked to volumetric pumping; and if special economic incentives are offered to those who use the treated wastewater, instead of freshwater, for many industrial purposes.

Chapter 12
Conclusions and Areas for Future Research

Abstract This chapter briefly discusses the findings and outcomes of the study on wastewater treatment and reuse that are covered in various chapters, vis-à-vis the key areas covered. It also discusses the future areas of research that would help design efficient wastewater treatment systems. It includes the key un-answered questions on the performance of the commonly used treatment systems in the country; performance of the technologies of removing complex compounds; performance of nature-based wastewater treatment system solutions; and, optimum level of performance of wastewater treatment systems that produce highest impacts in terms of their social, economic and environmental benefits.

Keywords Nature-based solutions · Environmental benefits · Social benefits · Economic value · Soil aquifer treatment · Constructed wetland

12.1 Summary of Findings and Conclusions

The urban population in India, which stood at 377 million in 2011, is expected to grow by 404 million, to become 781 million, by 2050 (United Nations 2014). The past growth rate in urban population in India has been higher in large cities as compared to small towns (Kundu 2006). Many large Indian cities get their water supply from sources far away, at high cost (Kumar 2014). Large Indian cities depend more on surface water than on groundwater. But in many cities and towns, groundwater is also an important resource. Only a small percentage of the effluent generated by cities and towns in India gets treated to the accepted standards, and the rest flows into rivers, lakes and tanks, rendering them unfit for human consumption. Pollution of groundwater due to untreated sewage collected in lakes and ponds is a serious issue in cities. On the other hand, sanitation has not kept pace with water supply in the country. The public health costs of diseases caused by water contamination and poor sanitation conditions are enormous.

A large proportion of the untreated and partially treated sewage from cities and towns located in water-scarce regions is also directly used in the peri urban areas for watering certain cereal crops, leafy vegetables and forage crops (Scott, Faruqui and Raschid-Sally 2004). Though this activity produces wealth for the farmers in the short run, long term use of such wastewater (Amerasinghe et al. 2013) which does not meet reuse standards, cause environmental and public health hazards in the form of soil degradation, groundwater contamination, direct impact on the health of people who handle wastewater and contamination of crops produced out of it especially with heavy metals (Source: based on Bose et al. 2010; Rawat et al. 2009; Winrock International India and others 2006).

In this book, the institutional, legal and policy measures for the control of water pollution were analyzed for their effectiveness in managing the quality of aquatic resources. We also analyzed the quality of water in surface water bodies across India using secondary data from government agencies, particularly the Central Pollution Control Board and the Comptroller and Auditor General of India, to assess the effectiveness of pollution control measures. The data on water quality in rivers upstream and downstream of selected cities were analyzed to understand the changes in water quality parameters over time. The literature on the human health impacts of water pollution in India was reviewed, using case studies of pollution hotspots. We also studied the status of wastewater generation, collection and treatment in urban areas across the country to understand the gaps in wastewater treatment. Wastewater treatment systems were studied *vis-à-vis* the technologies used, the capacity and performance of the plants, and their capital and O&M costs per unit volume of wastewater treated. The factors influencing the environmental sustainability and economic viability of wastewater treatment in Indian conditions were identified. Finally, the future scale of investment in wastewater treatment plants and the potential for reuse and recycling of treated wastewater in agriculture, industry and commercial uses were projected.

12.2 Future Research Areas on Wastewater

Though wastewater treatment systems involving different technologies are in use in Indian cities, information and knowledge about the technical and economic performance of these systems is too limited to be used in the design of efficient and cost-effective wastewater treatment systems for different situations. This gap poses a major challenge to design of wastewater treatment systems that are economically viable and socially and ecologically sustainable. The challenge of inadequate knowledge and information are on various fronts. We will discuss them.

First, the technical performance of the land-based wastewater treatment systems such as anaerobic systems and aerobic treatment systems (Activated Sludge Process, Up-flow Anaerobic Sludge Blanket, Anaerobic Pond, etc.) is heavily influenced by the climatic conditions, especially temperature (Peña et al. 2002; Richard 2003). In India, there is wide spatial and temporal variation in climatic conditions from hot and hyper arid to cold and cold and humid; and with wide fluctuation in mean daily temperature from cold winter to hot summer in some regions in the north and north west. While aerobic treatment systems perform well in cold and humid climate (Richard 2003), hot and arid climate is suited to anaerobic treatment systems (Peña et al. 2002). Since many of the old treatment plants were not designed by taking cognizance of this factor, the data on their performance will not be of much use, if the systems have to be designed using the same technology in another area with different climatic conditions. For instance, data on the performance of an aerobic treatment plant (on the technical and economic performance) using Activated Sludge Process functioning in a hot region will not be of much use in ascertaining the performance levels, if the same technology has to be used in place with cold climate. Similarly, the data generated from the plant on the technical performance (in terms of reduction in BOD/COD, pH, microbial and heavy metal content, etc.) and operation and maintenance cost cannot be used for design of a plant using anaerobic treatment technology. It is important that long-term data on the working of the same technology in similar climatic setting are available for obtaining efficient plant designs.

Second: many of the treatment plants that are in use in India that do secondary treatment of wastewater were not designed for removing complex compounds that are potentially present in wastewater discharge from hospitals, effluent from industrial units producing drugs and pharmaceuticals, and fertilizer and pesticide residues that are present in agricultural runoff (CPCB 2013b; Economic Times 2016). Instead, they were mostly meant for removing bio-chemical waste present in sewage (Planning Commission, 2011). The advanced treatment systems that are capable of treating toxic contaminants are of private sector industries, and there is hardly any information on the technical and economic performance of these systems in the public domain.

Third: there are many nature-based solutions for wastewater treatment such as constructed wetlands (Vymazal and Březinová 2014; GIZ 2011) and Soil Aquifer Treatment (SAT) (Idelovitch et al. 2003). They are extensively used in many developed and developing countries around the world (GIZ 2011; Icekson-Tal et al. 2003). For instance, in Israel, the Dan Region Reclamation Project, the largest wastewater treatment and reclamation project in that country, reclaims 130 MCM of wastewater per year for unrestricted irrigation after soil aquifer treatment (SAT) (Icekson-Tal et al. 2003). Extensive monitoring of system performance and aquifer water quality is performed to keep an efficient and safe wastewater reuse system (Idelovitch et al. 2003). SAT is used in the cities of Phoenix and Scottsdale in the United States for higher level treated of secondary treated wastewater (Aharoni et al. 2011). The factors influencing the performance is SAT are climate, geological/soil strata and geohydrology (Harun 1007). However, in spite of having a few regions that are suitable for its adoption, by virtue of their unique geology/soil strata, geohydrology and climate (Kumar 2014), application of this technology is almost absent in India. Same is the case with constructed wetlands. Hence, little is known about their effectiveness.

Generating knowledge about the economics of wastewater treatment is also essential for arriving at right management decisions. Treated wastewater produces several direct economic, social and environmental benefits, whose quantum depend on the quality of the treated wastewater and the socio-economic and ecological settings which receive the treated wastewater (Kumar 2014; Ramos et al. 2019). Adequately treated wastewater generates larger economic, social and environmental benefits than poorly treated wastewater (Milieu, WRc and RPA undated). Tertiary treated and sometimes quaternary treated wastewater can be used for direct human consumption, and therefore will produce large social benefits, including public health benefits in areas where freshwater resources are scarce.

However, not all the benefits would increase in proportion to the degree of treatment. The 'benefits' that would experience incremental change and the quantum would depend on the socio-economic and ecological settings in which the treated wastewater is discharged, particularly the demand for water in various water using sectors and the ecological carrying capacity of the region (Ramos et al. 2019). On the other hand, higher degree of treatment can also result in disproportionately higher capital and operating costs (Milieu, WRc and RPA undated). Hence it is important to know the optimum level of treatment of the wastewater that would produce maximum benefits in a given socio-economic and ecological setting.

Hence the future research should focus on the following: long term performance assessment of the some of the more common wastewater treatment systems under different climatic conditions, to know the climatic settings under which individual technologies offer the best performance and the costs involved; generating knowledge on the technical and economic performance of advanced technologies that are tested elsewhere for their efficiency in removing complex compounds present in hospital waste, effluent from pharmaceutical industry and fertilizer and pesticide residues from agricultural runoff, under various Indian conditions, through piloting; and assessing the performance levels of some of the nature-based solutions for water treatment by piloting them in settings under which they are likely to be effective; and finally knowing the optimum level of treatment from a plant that can produce maximum positive impact in terms of economic, environmental and social benefits in a given socio-economic and ecological setting.

Appendix

See Tables A.1, A.2, A.3, A.4, A.5 and A.6.

© The Author(s), under exclusive license to Springer Nature Singapore Pte Ltd. 2020 69
M. D. Kumar and C. Tortajada, *Assessing Wastewater Management in India*,
SpringerBriefs in Water Science and Technology,
https://doi.org/10.1007/978-981-15-2396-0

Table A.1 Water quality of the Musi River at Hyderabad

Year	Musi U/S of Hyderabad, A.P.									Musi D/S of Hyderabad, A.P.								
	DO (mg/L)			BOD (mg/L)			Faecal coliform (MPN/100 ml)			DO (mg/L)			BOD (mg/L)			Faecal coliform (MPN/100 ml)		
	Max	Min	Avg	Max	Min	Avg	Max	Min	Avg	Max	Min	Avg	Max	Min	Avg	Max	Min	Avg
2007	7.1	6.8	7	2	1.1	1.5	0	0	0	2.6	0	1	51	15	21.4	550	72	289
2008	11	5.2	7.59	4.2	0.5	1.68	165	0	24	4	0	1.24	23	5	12.9	220	2	98
2009	10.4	5.2	7.4	2	0	1.1	955	0	105	1.9	0	0.9	19	4	10.9	1680	0	285
2010	8.7	6	7.2	6	0.5	2.2	350	0	161	5.6	0	2.1	28	5	12.9	350	0	113
2011	13.8	4.1	6.9	7.5	0.5	2.8	1600	6	379	2.8	0.6	1.3	76	36.3	49.9	1600	11	702
2012	14.9	0	5.3	30	0	11.3	900	170	316	3.4	0	0.9	82	32	52.6	1600	350	1225
2013	6.7	6.1	6.4	7	7	7	280	170	225	0	0	0	75	50	62.5	1600	1600	1600

Source Central Pollution Control Board (2013a) & https://cpcb.nic.in/NWMP-data-2014

Table A.2 Water quality of the Godavari River at Nashik, Maharashtra

Year	Godavari U/S of Gangapur Dam, Nasik, Maharashtra									Godavari D/S of Nasik, Maharashtra								
	DO (mg/L)			BOD (mg/L)			Faecal coliform (MPN/100 ml)			DO (mg/L)			BOD (mg/L)			Faecal coliform (MPN/100 ml)		
	Max	Min	Avg	Max	Min	Avg	Max	Min	Avg	Max	Min	Avg	Max	Min	Avg	Max	Min	Avg
2007	7.5	5.9	6.6	9	4	5.8	13	0	5	5.9	3.2	4.8	36	7.2	15	250	7	59
2008	9.9	6.2	6.9	7	4	4.4	9	0	3	8.9	4.1	5.8	18	4.8	8.8	34	10	19
2009	6.7	5.2	5.8	12	3	5.9	14	2	5	6.9	4.4	5.3	14	4.5	9.2	25	12	17
2010	7.5	6.3	6.8	5	3	3.8	70	2	10	5.8	1.8	5	35	4.5	13.5	80	8	22
2011	6.9	5.8	6.5	9	2	4.2	14	7	11	7	1	4.8	28	2.4	11.2	50	11	25
2012	6.9	3	6.2	4	3	3.1	21	8	12	6.4	3	5.3	12	3	6.5	140	13	34
2013	7	5.7	5.5	3	2	2.5	44	8	22	6.5	4.2	5.2	8	3.2	5.2	111	13	52
2014	7.1	4.3	5.7	4.2	2.2	3.2	60	11	31	6.5	3.2	4.9	6	2.8	3.6	170	13	89

Source Central Pollution Control Board (2013a) & https://cpcb.nic.in/NWMP-data-2014

Table A.3 Water quality of the Yamuna River at Delhi

Year	Yamuna at Wazirabad Bridge, Delhi									Yamuna at Okhla Bridge (inlet of Agra Canal), Delhi								
	DO (mg/L)			BOD (mg/L)			Faecal coliform (MPN/100 ml)			DO (mg/L)			BOD (mg/L)			Faecal coliform (MPN/100 ml)		
	Max	Min	Avg	Max	Min	Avg	Max	Min	Avg	Max	Min	Avg	Max	Min	Avg	Max	Min	Avg
2007	12.1	5.5	8.3	6	1	3	26,000	200	5617	7.8	7	5	3.6	1	1.9	11,500,000	130,000	3,757,583
2008	10.5	6.2	8.1	3	1	1.5	60,000	500	18,600	1.4	0	0.2	55	2	25.7	2,900,000	189,000	709,900
2009	10	4.5	7.3	6	1	2.3	43,000	600	16,117	0.6	0	0.1	33	7	14.5	6,600,000	330,000	3,094,546
2010	10.2	5	7.8	2	1	1.3	15,000	540	3948	2.7	0	0.3	38	3	16.8	24,000,000	32,000	3,541,091
2011	10.3	5	7.8	4	1	2.2	46,000	400	9125	3.5	0	0.5	26	3	13	210,000,000	430,000	42,000,000
2012	11.4	6.1	8.8	3	1	2	1,000,000	1100	165,922	3.9	0	2	40	4	22	20,000,000,000	100,000	1,900,000,000
2013	11.5	9.5	10.5	4	2	3	3300	1700	2500	0.7	0.7	0.7	19	15	17	3,500,000	2,200,000	2,850,000
2014	12.8	6.4	9.3	5	1	2.4	43,000	450	10,003	3	0.4	1.4	37	5	17.3	160,000,000	450,000	23,816,667

Source Central Pollution Control Board (2013a) & https://cpcb.nic.in/NWMP-data-2014

Table A.4 Water quality of the Yamuna River at Mathura

Year	Yamuna U/S of Mathura									Yamuna D/S of Mathura								
	DO (mg/L)			BOD (mg/L)			Faecal coliform (MPN/100 ml)			DO (mg/L)			BOD (mg/L)			Faecal coliform (MPN/100 ml)		
	Max	Min	Avg	Max	Min	Avg	Max	Min	Avg	Max	Min	Avg	Max	Min	Avg	Max	Min	Avg
2007	7.8	3.7	5.8	20	6	10.8	47,000	1400	17,550	7.6	4.2	5 8	15	4	9.3	240,000	8500	50,692
2008	6.7	2.6	4.9	17	2	8.9	1,150,000	3700	178,100	8.4	2.5	5.8	18	4	10.6	1,010,000	4100	311,109
2009	11.8	1.9	5.7	14	5	9.2	290,000	10,400	89,950	7.1	2.8	6	16	6	9.4	500,000	37,000	182,417
2010	8.8	0.8	4.4	14	2	8.1	450,000	4300	187,775	20	1.5	7	13	3	7.8	930,000	9300	268,692
2011	8	0	3.8	41	5	11.4	4,300,000	9000	718,000	7.2	2.8	4.8	17	5	8.8	15,000,000	23,000	2,374,000
2012	8.8	0.8	4.4	14	2	8.1	700,000	2000	226,625	20	1.5	7	13	3	7.8	1,000,000	200,000	500,000

Source Central Pollution Control Board (2013a) & https://cpcb.nic.in/NWMP-data-2014

Table A.5 Water quality of the Yamuna River at Agra

Year	Yamuna U/S of Agra									Yamuna D/S of Agra								
	DO (mg/L)			BOD (mg/L)			Faecal coliform (MPN/100 ml)			DO (mg/L)			BOD (mg/L)			Faecal coliform (MPN/100 ml)		
	Max	Min	Avg	Max	Min	Avg	Max	Min	Avg	Max	Min	Avg	Max	Min	Avg	Max	Min	Avg
2007	14.6	3.8	7	16	5	10.2	460,000	2700	70,475	9	0	4.9	33	5	16.7	15,000,000	13,100	1,483,675
2008	8.7	2.2	5.9	17	8.8	7.7	800,000	3200	205,718	15.2	2.1	5.3	24	7	14	1,170,000	65,000	459,364
2009	12.4	3.4	6.3	11	3	8.1	1,500,000	21,000	232,818	6.7	0	4.1	32	6	14.5	5,400,000	62,000	1,403,546
2010	8.2	2.6	5.4	10	2	6.8	930,000	2700	298,418	21.1	1.6	6	15	4	9.7	930,000	31,000	260,182
2011	7.3	2.1	4.3	11	3	7	700,000	17,000	284,900	6.3	0.2	3.1	29	4	11	46,000,000	23,000	6,380,100
2012	8.2	2.6	5.4	10	2	6.8				21.1	1.6	6	15	4	9.7	900,000	30,000	475,000

Source Central Pollution Control Board (2013a) & https://cpcb.nic.in/NWMP–data–2014

Table A.6 Water quality of the Ganga River at Varanasi

Year	Ganga U/S of Varanasi (Assighat), U.P.									Ganga D/S of Varanasi (Malviya Bridge), U.P.								
	DO (mg/L)			BOD (mg/L)			Faecal coliform (MPN/100 ml)			DO (mg/L)			BOD (mg/L)			Faecal coliform (MPN/100 ml)		
	Max	Min	Avg	Max	Min	Avg	Max	Min	Avg	Max	Min	Avg	Max	Min	Avg	Max	Min	Avg
2007	9	8.6	8	3.8	3.1	3.3	13,000	8000	10,000	7.9	6.4	7.1	14	10	11.9	94,000	8000	81,714
2008	8.7	8.3	8.5	3.8	3.2	3.4	13,000	11,000	11,667	7.9	6.5	7.3	12.4	9.2	10.5	110,000	18,000	74,500
2009	8.5	8.2	8.4	3.8	3.3	3.5	13,000	8000	10,444	7.9	6.5	7.2	12.8	1	9.4	90,000	46,000	74,444
2010	8.7	7.4	8	4.2	3.4	3.9	11,000	6300	8730	7.7	6.8	7.3	10.8	4.2	8.9	49,000	33,000	44,400
2011	7.8	7.5	7.6	4.2	3.7	3.9	8000	8000	8000	7.2	7	7.1	9.6	5.2	8	46,000	34,000	40,000
2012	8.6	7.3	7.9	3.5	2.8	3.2	8000	5000	6784	8.1	6.8	7.4	5.4	3.6	5	33,000	27,000	31,667
2013	8.8	6.9	8.4	3.2	2.9	3	8000	5000	5500	8.3	6.4	7.9	5.1	3.3	4.6	46,000	31,000	34,500
2014	9.2	7.3	8.3	3.3	2.4	2.9	3300	2200	2617	8.7	6.7	7.8	5	3.8	4.5	46,000	22,000	31,167

Source Central Pollution Cortrol Board (2013a) & https://cpcb.nic.in/NWMP-data-2014

References

Agarwal SC, Kumar S (2011) Industrial water demand in India: challenges and implications for water pricing. In: IDFC (ed) India infrastructure report, chap 18. Oxford University Press, New Delhi

Aharoni A, Guttman J, Cikurel H, Sharma S (2011) Guidelines for design, operation and maintenance of SAT (and hybrid SAT) systems, Switch Project, UNESCO-IHE, May 2011

Al-Rekabi WS, Qiang H, Qiang WW (2007) Improvements in wastewater treatment technology. Pak J Nutr 6(2):104–110

Amarasinghe UA, Shah T, Anand BK (2008) India's water supply and demand from 2025–2050: business-as-usual scenario and issues. International Water Management Institute, Colombo, Sri Lanka

Amerasinghe P, Bhardwaj RM, Scott C, Jella K, Marshall F (2013) Urban wastewater and agricultural reuse challenges in India. Research report 147. International Water Management Institute, Colombo, Sri Lanka

Bassi N, Kumar MD, Sharma A, Pardha-Saradhi P (2014) Status of wetlands in India: a review of extent, ecosystem benefits, threats and management strategies. J Hydrol Reg Stud 2:1–19

Bedi JK, Ghuman RS, Bhullar AS (2015) Health and economic impact of unsafe drinking water a study of Ludhiana. Econ Political Wkly 50(2):23–26

Census of India (2011) 2011 census data. Government of India, New Delhi

Central Ground Water Board (2014) Ground water year book 2013–14. Ministry of Water Resources, Faridabad

Central Pollution Control Board (1999) Water quality status and statistics, 1996–97. Central Pollution Control Board, New Delhi

Central Pollution Control Board (2005a) Status of sewage treatment in India. Central Pollution Control Board, New Delhi

Central Pollution Control Board (2005b) Water quality status and statistics 2004–05. Central Pollution Control Board, New Delhi

Central Pollution Control Board (2008) Guidelines for water quality management. Ministry of Environment and Forests, New Delhi

Central Pollution Control Board (2013a) Status of water quality in India: 2012. Monitoring of Indian national aquatic resources series MINARS/36/2013-14. Ministry of Environment and Forests, New Delhi

Central Pollution Control Board (2013b) Performance evaluation of sewage treatment plants under NRCD. Government of India, New Delhi

Central Pollution Control Board (2015) Inventorization of sewage treatment plants. Control of urban pollution series CUPS/2015. Government of India, New Delhi

Central Pollution Control Board (2016) CPCB bulletin, vol 1, July. New Delhi

© The Author(s), under exclusive license to Springer Nature Singapore Pte Ltd. 2020

M. D. Kumar and C. Tortajada, *Assessing Wastewater Management in India*,

SpringerBriefs in Water Science and Technology,

https://doi.org/10.1007/978-981-15-2396-0

Central Public Health and Environmental Engineering Organisation (2005) Status of water supply, sanitation and solid waste management in urban areas. Research study series no. 88. Central Pollution Control Board, New Delhi

Central Public Health and Environmental Engineering Organisation and Deutsche Gesellschaft für Internationale Zusammenarbeit (2016) Municipal solid waste management manual. Ministry of Urban Development, New Delhi

Central Water Commission (2017) Reassessment of water availability in India using space inputs. Basin Planning and Management Organisation, Central Water Commission, New Delhi

Centre for Science and Environment (2014) Decentralized waste water treatment and reuse: case studies of implementation on different scale—community, institutional and individual building. Centre for Science and Environment, New Delhi

Comptroller and Auditor General of India (2011) Water pollution in India. Report no. 21 of 2011–12. Government of India, New Delhi

Deutsche Gesellschaft für Internationale Zusammenarbeit (GIZ) (2011) Technology review of constructed wetlands subsurface flow constructed wetlands for greywater and domestic wastewater treatment. Deutsche Gesellschaft für Internationale Zusammenarbeit (GIZ) GmbH, Eschborn, Germany

Economic Times (2016) Treating wastewater with the help of modern technology. Economic Times, Oct 25. http://economictimes.indiatimes.comhttps://economictimes.indiatimes.com/treating-wastewater-with-the-help-of-modern-technology/toshibashow_dp/55050324.cms?utm_source=contentofinterest&utm_medium=text&utm_campaign=cppst

Elangovan TG (2010) Efforts of CMWSSB to reduce pollution in Chennai city waterways. In: Proceedings of the seminar on waterways in Chennai, 4–5 Mar 2010. Chennai Metropolitan Development Authority, Chennai

Esposito K, Tsuchihashi R, Anderson J, Sekstrom J (2005) The role of water reclamation in water resources management in the 21st century. Water Environment Federation, Alexandria, VA

Global Water Partnership and International Network of Basin Organizations (2009) A handbook for integrated water resources management in basins. Stockholm

Government of India (1974) Water (prevention and control of pollution) Act, 1974. Government of India, New Delhi

Government of India (1977) Water (prevention and control of pollution) Cess Act, 1977. Ministry of Law, Justice and Company Affairs, Government of India, New Delhi

Government of India (1988) Water (prevention and control of pollution) Amendment Act, 1988. Government of India, New Delhi

Government of India (1999) Integrated water resources development: a plan for action. Report of the national commission on integrated water resources development. Ministry of Water Resources, Government of India, New Delhi

Government of India (2003) Water (prevention and control of pollution) Cess (Amendment) Act, 2003. Ministry of Law and Justice, Government of India, New Delhi

Government of India (2017) Guidelines for Swachh Bharat Mission—urban. Ministry of Housing and Urban Affairs, Government of India, New Delhi

Grey D, Sadoff W (2007) Sink or swim? Water security for growth and development. Water Policy 9(6):545–571

Hanjra M, Drechsel P, Mateo-Sagasta J, Otoo M, Hernández-Sancho F (2015) Assessing the finance and economics of resource recovery and reuse solutions across scales. In: Drechsel P, Qadir M, Wichelns D (eds) Wastewater. Springer, Dordrecht, pp 113–136

Harun CM (2007) Analysis of multiple contaminants during soil aquifer treatment. MSc thesis no MWI-2007-18, UNESCO-IHE. Index

http://cpcbenvis.nic.in/water_quality_data.html. Accessed 5 Dec 2017

http://www.pollutionissues.com/Ho-Li/Laws-and-Regulations-United-States.html. Accessed 7 Dec 2017

Icekson-Tal N, Avraham O, Sack J, Cikurel H (2003) Water reuse in Israel—the Dan Region Project: evaluation of water quality and reliability of plant's operation. Water Sci Technol Water Supply 3:239–246

Idelovitch E (2003) Soil aquifer treatment: the long-term performance of the Dan Region Reclamation Project. The World Bank water week 2003. Washington, DC

Husain I, Husain J, Arif M (2014) Socio-economic effect on socially-deprived communities of developing drinking water quality problems in arid and semi-arid area of central Rajasthan. Proc Int Assoc Hydrol Sci 364:229–235

Hussain J, Husain I, Arif M (2013) Fluoride contamination in groundwater of central Rajasthan, India and its toxicity in rural habitants. Toxicol Environ Chem 95(6):1048–1055

IDFC (2011) India infrastructure report 2011: water: policy and performance for sustainable development. Oxford University Press, New Delhi

Kiziloglu M, Turan M, Sahin U, Angin I, Anapali O, Okuroglu M (2007) Effects of wastewater irrigation on soil and cabbage-plant (*Brassica oleracea* var. Capitate cv. yalova-1) chemical properties. J Plant Nutr Soil Sci 170(1):166–172

Kodandaram U, Rao PM, Shashi CA (1972) Characteristics of waste from three typical tanneries in Madras. In: Proceedings: symposium on treatment and disposal of tannery and slaughter house waste. Central Leather Research Institute, Madras

Korsgaard L, Schou J (2010) Economic valuation of aquatic ecosystems in developing countries. Water Policy 12:20–31

Kumar MD (2010) Managing water in river basins: hydrology, economics, and institutions. Oxford University Press, New Delhi

Kumar MD (2014) Thirsty cities: how Indian cities can meet their water needs. Oxford University Press, New Delhi

Kumar MD (2018a) Future water management: myths in Indian agriculture. In: Biswas AK, Tortajada C, Rohner P (eds) Assessing global water mega trends. Water resources development and management series. Springer, Singapore, pp 187–209

Kumar MD (2018b) Water security in India: scenarios for 2040. Paper presented at the belt and road initiative workshop on global water security organized by Lee Kwan Yew School of Public Policy, National University of Singapore, Beijing, 14–15 Oct 2018

Kundu A (2006) Estimating urban population and its size class distribution at regional level in the context of demand for water: Methodological issues. Draft prepared for the IWMI-CPWF project on "Strategic Analysis of National River Linking Project of India"

Lok Sabha (2018) Reported cases and deaths by water-borne diseases in India. Lok Sabha, New Delhi

Milieu, WRc, Risk & Policy Analysts Ltd (RPA) (undated) Environmental, economic and social impacts of the use of sewage sludge on land. Final report part II: report on options and impacts, DG ENV.G.4/ETU/2008/0076r. Accessed at https://ec.europa.eu/environment/archives/waste/sludge/pdf/part_ii_report.pdf

Ministry of Environment, Forest and Climate Change (2000) Municipal solid wastes (management and handling) rules. Ministry of Environment, Forest and Climate Change, New Delhi

Ministry of Environment, Forest and Climate Change (2006) Ganga action plan—significant difference to water quality. https://pib.gov.in/newsite/erelcontent.aspx?relid=18777. Accessed 10 Oct 2019

Ministry of Environment, Forest and Climate Change (2007) Conservation of wetlands in India: a profile (approach and guidelines). MoEF, Government of India, New Delhi

Ministry of Environment, Forest and Climate Change (2016a) Solid waste management rules. http://bbmp.gov.in/documents/10180/1920333/SWM-Rules-2016.pdf/27c6b5e4-5265-4aee-bff6-451f28202cc8, http://envfor.nic.in/division/water-pollution. Accessed 17 May 2017

Ministry of Environment, Forest and Climate Change (2016b) Plastic waste management rules, 2016. Ministry of Environment, Forest and Climate Change, New Delhi

Ministry of Environment, Forest and Climate Change (2016c) E-waste (management) rules, 2016. Ministry of Environment, Forest and Climate Change, New Delhi

Ministry of Environment, Forest and Climate Change (2016d) Bio-medical waste management rules, 2016. Ministry of Environment, Forest and Climate Change, New Delhi

Ministry of Environment, Forest and Climate Change (2016e) Construction and demolition waste management rules, 2016. Ministry of Environment, Forest and Climate Change, New Delhi

Ministry of Environment, Forest and Climate Change (2016f) Hazardous and other wastes (management and trans-boundary movement) rules. Ministry of Environment, Forest and Climate Change, New Delhi

Ministry of Statistics and Programme Implementation (2016a) Compendium of environment statistics. http://www.indiaenvironmentportal.org.in/files/file/compendium%20of%20environment%20statistics%202016.pdf. Accessed 4 May 2017

Ministry of Statistics and Programme Implementation (2016b) Swachhta status report, 2016. Government of India, New Delhi

Ministry of Urban Development (2008) National urban sanitation policy 2008. Government of India, New Delhi

Ministry of Urban Development (2017a) National Urban Faecal Sludge and Septage Management Policy, draft policy document. Government of India, New Delhi

Ministry of Urban Development (2017b) Swachh Survekshan (urban sanitation report) 2017. Government of India, New Delhi

Mohammad MJ, Ayadi M (2005) Forage yield and nutrient uptake as influenced by secondary treated wastewater. J Plant Nutr 27(2):351–365

Mondal NC, Saxena VK, Singh VS (2005) Impact of pollution due to tanneries on groundwater regime. Curr Sci 1988–1994

Mukherjee S, Shah Z, Kumar MD (2010) Sustaining urban water supplies in India: increasing role of large reservoirs. Water Resour Manag 24(10):2035–2055

Murray A, Ray I (2010) Wastewater for agriculture: a reuse-oriented planning model and its application in peri-urban China. Water Res 44:1667–1679

Murty MN, Kumar S (2011) Water pollution in India: an economic appraisal. In: IDFC (ed) India infrastructure report, chap 19. National Capital Region Planning Board. Sewerage sector in the national capital region. Oxford University Press, New Delhi. http://ncrpb.nic.in/sewerage.html. Accessed 9 July 2017

National Capital Region Planning Board (2005) Regional plan 2021, national capital region. National Capital Region Planning Board, Ministry of Urban Development, Government of India, New Delhi

National Sample Survey Office (2014) Drinking water, sanitation, hygiene and housing condition in India. 69th round, July 2012–Dec 2012. Government of India, New Delhi

Nickson R, Sengupta C, Mitra P, Dave SN, Banerjee AK, Bhattacharya A, Basu S, Kakoti N, Moorthy NS, Wasuja M, Kumar M (2007) Current knowledge on the distribution of arsenic in groundwater in five states of India. J Environ Sci Health Part A 42(12):1707–1718

Organization for Economic Cooperation and Development (2008) OECD key environmental indicators. Paris, France

Pandey R (1997) Designing of pigouvian tax for pollution abatement in sugar industry. National Institute of Public Policy and Finance, New Delhi

Pandey P (2016) Japanese agency offer Rs 1000 crore to check pollution in Mula-Mutha. The Hindu, Mumbai

Pandey BW, Ranjan OJ, Srivastav A, Prasad AS (2017) Water pollution and its impact on human health: case study of Allahabad City, Uttar Pradesh. Int J Interdiscip Res Sci Soc Cult 3(1):14–25

Pant S (2018) GMDE draws GMDA draws up plan to reuse treated sewage water, lays pipes. Times of India, City Edition, Gurgaon, Haryana, India

Parker H, Oates N (2016) How do healthy rivers benefit society? A review of the evidence. Overseas Development Institute, Feb 2016. Accessed 17 Oct 2019. https://www.odi.org/sites/odi.org.uk/files/odi-assets/publications-opinion-files/10270.pdf

Peña MR, Madera CA, Mara DD (2002) Feasibility of waste stabilization pond technology in small municipalities of Colombia. Water Sci Technol 45(1):1–8

Planning Commission (2011) Report of the working group on urban and industrial water supply and sanitation for the twelfth five-year plan (2012–2017). Submitted to the steering group on water sector. Planning Commission, New Delhi

Press Information Bureau (2013) Merger of national lake conservation plan and national wetlands conservation programme into a new scheme. Cabinet Committee on Economic Affairs, Government of India, New Delhi

PwC (2016) Closing the water loop: reuse of treated wastewater in urban India. Knowledge paper. https://www.pwc.in

Ramos AV, Gonzalez ENA, Echeverri GT, Moreno LS, Díaz Jiménez L, Hernández SC (2019) Potential uses of treated municipal wastewater in a semiarid region of Mexico. Sustainability 11:2217

Rawat M, Ramanathan A, Subramanian V (2009) Quantification and distribution of heavy metals from small-scale industrial areas of Kanpur city, India. J Hazard Mater 172(2–3):1145–1149

Richard M (2003) Activated sludge microbiology problems and their control. Presented at the 20th annual USEPA national operator trainers conference, Buffalo, NY, 8 June 2003

Rosenqvist H, Aronsson P, Hasselgren K, Perttu K (1997) Economics of using municipal wastewater irrigation of willow coppice crops. Biomass Bioenergy 12(1):1–8

Roy C (2012) A study on environmental compliance of Indian leather industry & its far-reaching impact on leather exports. Munich Personal RePEc Archive. https://mpra.ub.uni-muenchen.de/41386/1/A_Study_on_the_Environmental_Compliance_on_Indian_Leather_Industry-Ready.pdf

Sahasranaman M, Ganguly A (2018). Wastewater treatment for water security in India. IRAP occasional paper no. 13. Institute for Resource Analysis and Policy, Hyderabad, India

Scott CA, Faruqui NI, Raschid-Sally L (2004) Wastewater use in irrigated agriculture confronting the livelihood and environmental realities. CABI Publishing, CAB International, Wallingford, Oxfordshire, United Kingdom

Seth BL (2011) Water pollution hot spots identified in Indian rivers. Down To Earth, 29 Nov. http://www.downtoearth.org.in/news/water-pollution-hot-spots-identified-in-indian-rivers-34526. Accessed 16 May 2017

Srikanth R (2009) Challenges of sustainable water quality management in rural India. Curr Sci 97(3):317–325

Telangana Today (2017) Water board to sell treated domestic waste water, 3 Mar. https://telanganatoday.com/water-board-to-sell-treated-domestic-waste-water. Accessed 9 July 2017

Tewari S, Bapat R (2016) The great Indian water walkathon. IndiaSpend, 21 June. https://archive.indiaspend.com/cover-story/the-great-indian-water-walkathon-26344

Tripathi B (2018) Diarrhoea took more lives than any other water-borne disease in India. IndiaSpend, New Delhi

United Nations, Department of Economic and Social Affairs, Population Division (2014) World urbanization prospects: the 2014 revision. Highlights (ST/ESA/SER.A/352)

Urban Design Research Institute (n.d.) Sewage treatment plant. http://www.mumbaidp24seven.in/reference/stp.pdf. Accessed 10 July 2017

US EPA (United States Environmental Protection Agency) (2003) Control of pathogens and vector attraction in sewage sludge; 40 CFR part 503. Environmental Protection Agency, Cincinnati, p 45268

Vymazal T, Březinová T (2014) Long term treatment performance of constructed wetlands for wastewater treatment in mountain areas: four case studies from the Czech Republic. Ecol Eng 71:578–583

Wani D, Pandit AK, Kamili ZN (2013) Microbial assessment and effect of seasonal change on the removal efficiency of FAB based sewage treatment plant. J Environ Eng Ecol Sci 2:1–3

WaterAid (2016a) An assessment of faecal sludge management (FSM) policies and programmes at the national and select states levels. WaterAid, New Delhi

WaterAid (2016b) Drinking water quality in rural India: issues and approaches. Background paper. WaterAid, New Delhi

Water Resources Information System of India. http://www.india-wris.nrsc.gov.in/wrpinfo/index. php?title=India%27s_Water_Wealth. Accessed 4 May 2017

Winrock International India, Institute for Studies and Transformations, Jadavpur University, Department of Economics, Eco Friends, Spatial Decisions, Youth for Unity and Voluntary Action (YUVA) (2006) Urban wastewater: livelihoods, health and environmental impacts in India. Research report submitted to comprehensive assessment of water management in agriculture. Winrock International India, New Delhi, India, 160 pp. Available at www.iwmi.cgiar.org/ Assessment/files_new/research_projects/Urban%20Wastewater-Full_Report.pdf. Accessed 16 Nov 2012

World Bank (2008) Review of effectiveness of rural supply schemes in India. World Bank, New Delhi

WWAP (United Nations World Water Assessment Programme) (2017) The United Nations world water development report 2017. Wastewater: the untapped resource. UNESCO, Paris

Index

A

Agriculture, 31, 33, 48, 51, 52, 54, 56, 59–63, 66

Anaerobic lagoons, 37, 38, 41

Aquatic resources, 4, 8, 11, 16, 66

Aquifers, 2, 13, 67

Areas for future research, 65

B

Bio-chemical waste, 67

Black water, 21, 40

C

Cases of reuse of treated wastewater, 31

Central Ground Water Board (CGWB), 2, 5

Central Pollution Control Board (CPCB) Pollution Control Committee, 9

Central Public Health and Environmental Engineering Organization (CPHEEO), 10

Central Water Commission, 56

Chennai Metropolitan Water and Sewerage Board (CMWSB), 32

Class I cities, 20, 22, 37, 54–56, 61

Class II cities, 17, 18, 22, 54

Climate, 1, 9–11, 37, 41, 52, 66, 67

Compliance with pollution control norms, 14

Conventional or centralized wastewater treatment, 36

D

Decentralized systems for wastewater treatment:performance and costs, 39

Drinking water, 2, 9, 15, 24, 27, 29, 32

E

Eco-development, 9

Ecological health, 54, 57, 63

Economic capital and O&M costs, 42, 43, 66

Economic viability, 49, 50, 59, 62, 66

Effectiveness of wastewater collection and treatment systems, 17

Environment, 2, 4, 9–11, 13–15, 29, 41, 50, 53

Environmental benefits, 62, 65, 67

Environmental management services healthy aquatic life, water for recreation, 59

improved quality of river water, 59

improved soil quality, 59

Environmental sustainability, 49, 50, 66

Environmental sustainability versus economic viability, 50

F

Fertilizer and pesticides, 14, 67, 68

Freshwater, 2, 3, 14, 16, 17, 31, 51, 54, 59, 61, 63, 68

Future growth of treatment plants and wastewater reuse, 53

Future investments for wastewater treatment, 58

Future market for treated wastewater, 60

Future research areas on wastewater, 66

Future treated wastewater market potential, 3, 4, 53, 59, 60

Future water demand, 60

Future water supplies, 60

G
General profile of plants and locations, 43
Geohydrology, 67
Global environmental monitoring system, 8
Global water partnership and international
    network of basin organizations, 3
Groundwater, 1, 2, 5, 8, 13–16, 21, 24,
    29–31, 40, 51, 60, 63, 65, 66
Growth of treatment plants and reuse of
    treated wastewater, 53
Guidelines for water quality management,
    11

H
Health impacts of water pollution and con-
    tamination, 23

I
India
    Agra, 16, 72, 74
    Assam, 21, 51
    Bangalore, 32, 51, 54
    Bihar, 21
    Chandigarh, 21
    Delhi, 16, 17, 19, 21, 22, 32, 43–48, 51,
        54, 61
    Haryana, 14, 20, 31
    Himachal Pradesh, 21
    Hyderabad, 16, 22, 32, 43–48, 51, 70
    Karnataka, 16, 19, 22
    Kerala, 21, 51, 54
    Ludhiana, 24, 30
    Maharashtra, 19, 20, 22, 45, 47, 49, 71
    Mathura, 16, 73
    Mumbai, 17, 18
    Punjab, 14, 20, 24
    Tamil Nadu, 20, 30, 49, 51
    Uttar Pradesh, 19, 20, 29, 30, 49
    Varanasi, 15, 16, 75
Industrialization, 2, 29
Infrastructure development finance com-
    pany, 13, 14
International Water Management Institute
    (IWMI), 1
Investment, 3, 4, 9, 27, 33, 43, 47, 53–58, 60,
    61, 63, 66
Irrigation systems, 52

L
Land prices, 50
Legal and institutional regime for water
    quality management, 7, 8

M
Major crops
    paddy, 60
    sugarcane, 60
    wheat, 60
Market for treated wastewater in India, 59
Ministry of Environment, Forest and Climate
    Change, 9–11
Ministry of Statistics and Programme Imple-
    mentation, 11, 21
Ministry of Urban Development, 10, 11, 15,
    16, 21
Ministry of Water Resources, 1
Monitoring of Indian National Aquatic
    Resources Series (MINARS), 11

N
National Capital Region Planning Board
    (NCRPB), 32, 33
National River Conservation Directorate
    (NRCD), 9, 20, 37
National River Conservation Directorate's
    National Lake Conservation Plan, 9
National River Conservation Plan (NRCP),
    9, 14, 15
National Urban Faecal Sludge and Septage
    Management Policy, 10
National Urban Sanitation Policy, 9–11
National Wetlands Conservation Pro-
    gramme, 9

O
Organic pollutants, 12, 24

P
Performance of wastewater treatment sys-
    tems, 43
Pharmaceuticals, 67, 68
Plants and locations, 43
Polluter pay's principle, 9
Pollution control issues, 12
Pollution control norms, 12–14, 16, 54
Population growth, 17, 53
Prevention and control of pollution, 7, 11, 12
Public health, 1, 3, 4, 9–11, 21, 23, 24, 30,
    51, 65, 66, 68

Public health impacts of water contamina-
    tion in India, 23

R
Recycling
    recycling and resource recovery, 9
Reuse of treated wastewater, 4, 31, 51, 58
Rivers
    Allahabad, 29
    Ganges, 30
    Godavari, 45, 71
    Musi, 32, 45, 70
    Mutha, 45, 47
    Noyal, 51
Rural areas, 2, 13, 21, 35, 39, 40, 54

S
Sanitation, 1–3, 7, 9–11, 15, 16, 21, 23, 29,
    30, 39, 40, 42, 50, 65
Sanitation access in urban India, 21
Septic tank systems
    pit latrines, 11, 35, 40
    systems generating faecal sludge, 11
Sewage treatment capacity, 18
Sewage treatment plants:performance and
    costs, 4, 9, 15, 18, 38, 39, 43
Sewerage system, 2, 7, 18, 20, 22, 48
Social benefits, 49, 59, 68
Soil aquifer treatment, 40, 50, 67
Solid waste management rules, 10, 11
Status of water quality in India and com-
    pliance with pollution control norms,
    13
Stormwater, 3, 32

T
Technology
    advanced, 36, 37, 68
    conventional, 37, 38
Temperature, 25, 27, 40, 50, 66
Treated wastewater, 3, 15, 22, 31–33, 40,
    43–45, 48, 50–54, 60, 61, 63
Treatment plants, 5, 17, 19, 32, 36, 43–45,
    47–50, 52, 53, 56–58, 61, 63, 66, 67
Treatment technologies used in sewage treat-
    ment plants: performance and costs,
    37

U
United Nations World Water Assessment
    Programme (WWAP), 7

United States
    Phoenix, 67
    Scottsdale, 67
Urban areas, 2, 3, 10, 14, 17, 21, 31, 35, 36,
    39, 49, 61, 65, 66

W
Waste stabilization pond, 38
Wastewater
    agricultural uses, 56, 63
    benefits treated wastewater, 59
    domestic wastewater, 18, 22, 32, 35
    economic, environmental and social
        aspects, 49, 59, 68
    horticultural purposes for wastewater, 22
    industrial wastewater, 18, 35
    treated wastewater markets, 3, 60
Wastewater treatment technologies and
    costs, 35–41
    conventional treatment, 9
    costs of wastewater treatment, 35
    decentralized wastewater treatment, 40
    economics of wastewater treatment, 4,
        50, 67
    nature-based treatment, 67, 68
    secondary treatment, 48, 56
    tertiary treatment, 57, 61, 63
    wastewater collection, 17–22
    wastewater generation, 56
    wastewater management standards (envi-
        ronmental management standards), 54
    wastewater treatment capacity utiliza-
        tion, 17
    wastewater reuse policies, 63
Wastewater generation and collection in
    India's urban areas, 17
Water demand, 1–3, 33, 51, 53, 60–62
Water infrastructure, 49
Water management, 1, 3, 5, 13
Water pollution and impacts on public
    health, 2, 4, 7, 8, 11, 12, 14, 22, 24,
    29, 30, 66
Water quality monitoring in India, 8
    Water quality management, 5, 7, 8, 11
    water quality monitoring, 7, 8, 12, 16, 43,
        47, 49
Water quality status in India, 13
Water Resources Information System, 1
Water reuse, 3
Water rich regions, 54–57, 60–63
Water security, 2, 3, 16
Water stress, 2, 53
Water treatment, 8, 68

Water-borne diseases
  amoebiasis, 24
  cholera, 24, 25, 28, 30
  dengue, 24, 27
  diarrhoea, 24, 28, 30
  dysentery, 24–26, 30
  jaundice, 24
  malaria, 24, 27
  paratyphoid fever, 24
  typhoid, 24, 25, 28, 30

Water quality, 1, 4, 7–9, 11–16, 22, 24, 43,
  47, 48, 51, 54, 66, 67, 70–75
Working of the wastewater treatment plants,
  45
World Bank, 2, 3

Y
Yamuna Action Plan, 8, 20

Printed in the United States
By Bookmasters